Eco-design of Marine Infrastructures

Series Editor
Françoise Gaill

Eco-design of Marine Infrastructures

Towards Ecologically-informed Coastal and Ocean Development

Sylvain Pioch
Jean-Claude Souche

iSTE

WILEY

First published 2021 in Great Britain and the United States by ISTE Ltd and John Wiley & Sons, Inc.

ISTE Ltd
27-37 St George's Road
London SW19 4EU
UK

www.iste.co.uk

John Wiley & Sons, Inc.
111 River Street
Hoboken, NJ 07030
USA

www.wiley.com

Library of Congress Control Number: 2021940543

British Library Cataloguing-in-Publication Data
A CIP record for this book is available from the British Library
ISBN 978-1-78630-711-8

FSC
www.fsc.org
MIX
Paper from responsible sources
FSC® C013604

Contents

Foreword

Grown-ups never understand anything by themselves, and it is tiresome for children to be always and forever explaining things to them.

Antoine de Saint-Exupéry

Our planet and our ecosystems are in danger and we must act. Of course, we hear this message every day and most of us are now aware of it. However, it is not easy to exactly identify the different challenges ahead, and it is even less easy to act accordingly.

Intended for a wide audience, this book can help provide answers. It presents, through concrete examples and testimonies, an exhaustive state of the art allowing everyone to better understand marine eco-design and the issues it addresses. It also proposes a methodology for acting differently. Although this book is primarily intended for technicians, engineers, scientists and students, it may be of interest to anyone who is curious to see how we can "develop" by taking inspiration from nature. For this book is not only the story of two men of art, it is also the work of two marine enthusiasts who, for more than 30 years, have been working for the preservation of the seabed; passionate people who have spent hundreds of hours in the water observing, marveling at marine life and trying to understand the combination of elements and the consequences on biodiversity. I have shared this passion with them for many years.

Through this book, Sylvain and Jean-Claude, who are great professionals and long-time friends, will share their universe with you. Through concrete experiences, you will discover the marine world, see through their eyes the underwater biotope and how to preserve it and perhaps feel its mysteries in order to reconcile what seems irreconcilable: the human impact of a maritime infrastructure and the preservation of biodiversity. Sylvain Pioch combines both a sixth sense of this wilderness and an exceptional scientific knowledge. His ability to understand and predict the behavior of fish and marine life will always amaze me. He is, at the same time, a renowned professor, a researcher and a talented designer, recognized throughout the world. Jean-Claude Souche is the one who makes it possible to transform concepts into sustainable developments and infrastructures. Today, Jean-Claude is a professor at IMT Mines Alès, a French engineering school, where he heads the civil engineering and sustainable building department. His international experience in marine works, his operational background as an engineer and doctor and his unwavering will to move forward make him a valuable person. With their common conviction, on a professional level, Sylvain and Jean-Claude are incredibly complementary. In this book, they have produced a thorough work on the eco-design of marine infrastructures, which for them is definitive work. This book highlights one of the great challenges of our century, that of preserving marine biodiversity through our land use planning policy and the construction of coastal and port infrastructures. The protection of the environment in terms of maritime development is no longer based on simple compensatory measures, but must, by definition for any project, preserve and promote the development of life and its diversity. I am deeply convinced that project owners, engineers, scientists and contractors can and must play a major role in promoting biodiversity and thus protecting the marine environment. One of the major challenges to come will be to know how to create industrial policies that are economically efficient and respectful of our environment. It is no longer a question of opposing environment and economy or human technology and nature, but, on the contrary, of reconciling them. This requires skill and conviction. Even though science and technology are not the only solution to the problems of today's world, they provide technical innovation that can lead to changes in human behavior. This is the goal of this book, to allow as many people as possible to understand the stakes, to act intelligently and, finally, to think differently.

Also, throughout these pages, I urge you to think about what we will leave to future generations, I urge you to become the children of the *Little Prince* because it is possible to change the world, provided we all change, provided we first change ourselves.

Enjoy reading this book.

Régis DUMAY
Deputy Managing Director, Egis

Preface

> The distance of man's emancipation from the sea is equal to the distance of our cells from the composition of sea water.
>
> Loren Eiseley

The purpose of this book is to strengthen the path towards a coastal maritime management where civil maritime engineering is intimately linked with environmental engineering, within a socio-ecosystem where humanity is an integral part of nature. The multiple consequences of the mistreatment of nature by a denatured human will not be discussed in this book. Indeed, it seems to us that the links between the artificialization of the seabed, climate change, pollution, the introduction of invasive species or the overexploitation of natural resources with a deregulated, predatory and irresponsible anthropic activity for the future of ecosystems and the survival of humanity (on our unique Earth) are obvious (IPBES 2019). Neither are we prophets, as the concepts discussed here have already been the subject of modern works (Belknap *et al.* 1967; Falque 1972; Tarlet 1977) or of older, empirical findings, where humans have also illustrated themselves in their capacity for positive interactions with nature (McHarg 1969; Lassus 2002).

The geographer and planner McHarg (1969)[1] detected that in our current modern societies (since the second half of the 20th century), technicists, industrialists and urbanists have an attitude to the human that is dissociated, pre-Copernican and dominant towards nature. The source of this segregative reflection is that we have been fed by an ancient instinct of revenge towards nature, born of a 1,000-year-old resentment of having held so little influence before nature. Psychoanalysts would

1 The work of McHarg in the area of landscape architecture, gathered in his famous *Design with Nature*, served as a basis for our extrapolation to the submerged, underwater marine domain.

call this a "complex of cultural inferiority, with perverse aggressive tendency", that is, without consideration, nor empathy, nor feeling for the tormented object. This conception of nature, which is the subject of our predation, would satisfy the desire for primacy buried deep within the human being, for a long time inoffensive since impossible to achieve technically. How can we make this resentment, which we have historically inherited as a consequence of our environment, null and void? How can we prevent it from poisoning the objective of survival and evolution of a human who can now "stand up among the other forms of life" (McHarg 1969)? The expression of our work on eco-design is rooted in our enthusiasm to assert our talents as creators, rather than those of destroyers who are less worthy to represent responsible humans, the managers of their environment and thus of their future.

This exercise is, moreover, made difficult by the Western conception founded on an anthropocentrism disassociated with nature, notably spiritually (Berque 1986). The oriental approach, for example, the Japanese approach using Tao, Shinto or Zen, has sometimes ignored the human as an individual to focus on the human within nature (the garden being the metaphysical symbol par excellence).

In short, two reverse postulates exist: in the West, the human at the expense of nature, and, in the East, nature at the expense of the individual human. The third view would be that of a balance, which does not mean a fusion, where the human is considered as an individual, rather than as a species, within nature.

To date, however, this way has not been expressed in human "works" presented on the maritime domain (principally the submerged part), which have never taken into account natural facts in their intrinsic conception. It is the human against nature, which is understood in maritime engineering as a vocabulary of work or technique: works of defense against the sea, breakwaters, dikes, wave-breaking walls, seawalls, dredging, etc.

On the contrary, land constructions have long been based on a local empirism (the vernacular), allowing humans to observe nature and to settle there harmoniously. The low stone walls follow the curves of hillsides where the peasantry, better than any other profession, know by observation how to exploit and manage the land. There are also our medieval "circulade" Mediterranean villages, where the air circulates wonderfully and naturally refreshes the shaded alleys, offering nesting boxes to swallows and swifts feasting on mosquitoes near the houses. Contrast this with the modern suburbs on the outskirts of these same villages which are asphyxiated, overheated by increasing heat waves and have often disfigured the harmony of the landscape.

We are convinced that the construction of structures must be sensitive to the laws and needs of nature, to ecosystems, to materials and forms adapted to human needs and to the beauty of life, and thus offer sustainable achievements. Eco-design will therefore be adapted to the place and will bring long-term benefits to humans and nature. It is based on ecology, from the Greek *oikos*, or house, that is, the science of the dwelling, an obvious prerequisite for any development whose objective is to arrange with order (and according to the rules of ecology, of the human in nature) human settlements, with a view to sustainable and desirable development: managing life to ensure our survival.

The temptation and the drift towards a cosmetic nature, a simple green washing, is always present, but a detailed knowledge of the natural functioning and of the typical ecosystem for each site and each project, targeting an integration between the ecological and aesthetic landscape, as well as an ecological follow-up of the developments, are the guarantees to keep a good course. We will have at least tried to advance the notion that the human can play the role of positive creator for their environment, improving the biosphere with new symbioses of humans in nature.

In this work, we propose to focus on marine structures. Indeed, coastal structures, although not specifically designed for this purpose, generate new biotopes that are particularly attractive for coastal species at the juvenile stage: for example, 30–109 times more juveniles are welcomed on dikes and harbors than on natural rocky habitats in the western French Mediterranean (Pastor 2008).

However, these ecological potentialities are ignored, or at best incidentally recognized and very rarely enhanced by specific eco-designed and nature-inspired infrastructures. Marine works are generally designed with regard to functional, technical, economic or hydro-sedimentary marine environment considerations, not as supports to maintain or increase marine biodiversity.

The objective is now to develop their functional aspects from an ecological point of view so that the structure becomes a proactive element for the environment. It becomes part of a dynamic ecosystem by creating habitats and ecological functions: shelter for juveniles, feeding areas, habitats for fixed fauna and flora, etc.

This book first offers the reader two chapters related to the developments in the fields of environmental regulation and maritime civil engineering, increasingly expressing a social expectation towards the prefix "eco-". Indeed, every planner must design a facility in response to a functional and technical need, meeting regulatory standards.

However, beyond this "classic" approach to marine development projects, the eco-design approach presented in Chapter 3 allows environmental impacts to be taken into account from the technical definition of the works. The authors illustrate this approach in Chapter 4 with practical examples that they have dealt with, complemented by feedback from projects carried out according to these principles and attempting to avoid running the risk of justifying avoidable projects by their "green" appearance. It is always preferable to abandon a project if its negative effects on the environment are unavoidable, because offset is a Trojan horse for the development that is thus facilitated, as denounced by Firth *et al.* (2020): "Greening of grey infrastructure should not be used as a Trojan horse to facilitate coastal development."

Sylvain Pioch
Jean-Claude Souche
July 2021

Acknowledgments

This book is dedicated to our families: Élise, Guilhem, Céleste, Raphaëlle and Julianne, and Sophie, Juliette, Rémi, Éloïse and Alexis, who are the driving force of our lives. We would also like to thank all those who, through their lives, actions and convictions, seek to build a fairer and more beautiful world where humankind will be able to consider all the other forms of life on the planet of which it is only the host.

This book is the compilation of 30 years of professional practice acquired as an environmental engineer in planning and impact studies (S. Pioch) and as a civil engineer (J.-C. Souche), then as lecturers and researchers in University Montpellier 3 and Mine Telecom Institute of Alès or simply as divers.

Beyond our common passion for the sea, it seemed unavoidable to merge our experiences as the need to design and build marine structures differently is urgent. It is also with modesty that we propose this work to readers, without claiming it to be exhaustive or the absolute truth. We simply provide a testimony and methodological tracks that we have tested in the field, with actors involved in development.

We would like to thank Régis Dumay, Marie Salgues, Jean Bougis, Philippe Saussol and Jean-Marie Miossec for their active help in the elaboration of this book, as well as the company Beuchat for the diving equipment. A special thought for Françoise Gaill for her constant support. We would also like to thank all the contributors who shared their testimonies with us in order to make this book a moment of sharing of experiences in France and abroad, which we hope the readers will appreciate:

– Alexandra Agostini, Anne Rioux, Aurore Léocadie, Osanne Paireau, Katherine A. Dafforn (Australia), Louise Firth (UK) and Shimrit Perkol-Finkel (Israel);

– André Grosset, Bernard Sigros, Cyril Giraudel, Cyrille Taioni, Etienne Cunge, Fabrice Javel, Frédéric Martarèche, Jean-Louis Gaziello, Jean-Michel Cathala, Michel Fons, Patrick Guiraud, Pierre Roy, Romaric Vicente, Marcel Stive (Netherlands), Jean-Luc Nguyen (Monaco) and Richard Spieler (USA).

May this modest work inspire us to think differently and to eco-design marine works, for the mutual benefit of the sea and humankind.

1

Principles and Genesis of Maritime Eco-design

The global ecological crisis is concomitant with the Anthropocene, this new geological era in which humans have become the central actors of pressures on the planet (Crutzen 2006). Indeed, the recent IPBES report (IPBES 2019) makes a damning assessment of the state of biodiversity since the beginning of the industrial era two centuries ago: 75% of the land has been altered by humans (one-third of the land consuming three-quarters of the available water resources is agricultural), 66% of the oceans are threatened by humans and more than 85% of wetlands have been destroyed. It is therefore our actions on this unique planet that are holding back our own future. The logic of this observation would lead us to stop, or at least to slow down, the well-known causes of this disaster (in decreasing order): (1) artificialization and land use; (2) resource exploitation (fishing, forestry, etc.); (3) climate change; (4) pollution (plastics, chemical residues, etc.); and (5) invasive species (IPBES 2019). Thus, artificialization and land use would be our main problem. In addition to agriculture, it is the issue of urbanization and its consequences (cities, ports, mines, industries and roads) that is the most important because it leads to an artificialization of environments that is difficult to reverse. If we look at the forecasts, we can see not only a continuity but also an acceleration of global urbanization, both on the continents and at sea. No less than 60,000 billion US dollars will be invested in infrastructure between 2019 and 2040 (in the 56 countries representing 88% of the world's GDP (Global Infrastructure Hub and Oxford Economics 2017)), more than 1.2 million km^2 will be urbanized by 2030, in just 10 years, that is, an increase of 185% compared to 1970–2000 (Seto *et al.* 2012), and 3–4.7 million km^2 of roads will be created by 2050, an increase of 25% compared to the current annual rate (Meijer *et al.* 2018).

This phenomenon is particularly prevalent in coastal areas, as 8 of the 10 global megacities are located on shores, such as Lagos, Tokyo, Jakarta and New York. The global maritime infrastructure footprint was approximately 32,000 km^2 in 2018 (Bugnot *et al.* 2021). It is expected to reach 39,400 km^2 by 2028, a territory equivalent to the state of Bhutan! However, the area of seascapes impacted by these structures was also estimated to be between 1 and 3.4 million km^2 in 2018, with an increase of 50–70% expected by 2028, which is comparable to the global extent of urbanized areas (estimated at 0.02–1.7% of the land mass (Bugnot *et al.* 2021)). On the French coast, the rate of occupation doubled between 1965 and 1980[1]. Between 2000 and 2006, no less than 6,809 ha were destroyed for the construction of harbors, dikes, embankments and other structures[2]. For example, in Hérault, a department in the Occitanie region, close to the Mediterranean Sea (southern France), the level of urbanization in 2015 on a narrow coastal strip of 15 km was close to 70%, with no less than three-quarters of the population concentrated there (DREAL Occitanie). In the international literature, we speak of a global "coastal squeeze". This phrase, introduced by Doody (2004), is based on the threat to coastal development caused by the dual effect of rising sea levels and the explosion of human activities. There is an incoherence between human and the needs of our planet, especially in the urgent matter of the artificialization of natural environments, notably coastal ones. Urbanization remains the result of a development that is still too predatory of space.

Faced with the enormous challenge of a renaturation of culture (Pelt 1977), for a livable future of humankind, it becomes crucial to improve the consideration of biodiversity in territorial planning projects. We will focus here on the potential for nature-friendly planning, trying to integrate its functional needs as a fully-fledged objective in the design of infrastructures.

1.1. Principles of maritime eco-design

The actions taken to allow for the "natural" environment (we will use this term here in relation to the word "ecosystem") for the operation of maritime works are varied. In the case of ports, these include the control and reduction of discharges, energy, sediment, waste and water management, environmental management plans (compliant since 2013), natural infrastructure master plans – including a natural heritage master plan – and Natura 2000 operators within the port perimeter (e.g. the Grand Port Maritime of Dunkirk).

1 MEDAM (2015). Database for the French Mediterranean Coast (Inventory and Impact of Land Reclamation).

2 *Ibid.*

At the same time, several tools designed to provide (voluntary) environmental certification are also available to project owners: AFAQ Clean Ports, ISO 14001, Blue Flag and the latest "Clean Port" certification active in biodiversity[3], in March 2018. These certifications are often accompanied by communication programs, such as the "Green Port"[4], "Year of Biodiversity" and "Port Biodiversity Index", or information panels and stands proposed by the ports' sustainable development departments, suppliers of "eco-" equipment or operators (Figure 1.1).

Figure 1.1. *Public presentation panel on port biodiversity in the port of Kernével, one of the first two ports in Brittany to be certified "Clean Ports Active in Biodiversity"[5] in 2018 (photo: ©APPB)*

Political actions or the dissemination of good environmental practices have been strengthened in recent years. At the international level, for example, we can cite the Working with Nature program (from the International Association of Ports and Canals, 2008), based in particular on the experience of Port 2000 in Le Havre and the numerous guides produced PIANC[6] (2011a, 2019, 2020); the World Harbour Project[7] (Steinberg *et al.* 2016), which brings together 15 ports around the issues of ecological engineering, nature-based solutions and the resilience of natural port

3 This AC J81-032 certification of March 2018 is an optional component that adds to the requirements of the "Clean Ports" CWA AC J 81-030 standard and is part of a prior European "Clean Ports" certification process.

4 http://www.greenport.com.

5 http://www.portsdebretagne.fr/actualites/deux-ports-bretons-certifies-ports-propres-actifs-en-biodiversite-roscoff-bloscon-et-kernevel/.

6 The World Association for Waterborne Transport Infrastructure (PIANC), the oldest international professional organization in the port field: https://www.pianc.org.

7 http://www.worldharbourproject.org.

environments; at the European level, the CWA 16987 (Clean Harbour Guidelines); and at the national level, the reflections initiated as part of the Grenelle mission on the "Port of the Future" (led by CEREMA[8]), or regional variations, such as for the major seaport of Marseilles and the "GIREL"[9] research program, which was carried out in 2010.

In spite of these virtuous impulses, in the field, during the first design phases of a project, the objectives are primarily to propose a structure that meets technical constraints (resistance, durability) with a controlled cost, aligned with socio-economic objectives that meet a functional need: a marina or a commercial port, an offshore wind turbine, a breakwater, an offshore wastewater treatment plant, etc. The environmental question is applied to justificatory and secondary considerations which are dealt with once the technical and socio-economic choices have been made, under regulatory "constraint" (Airoldi *et al.* 2021).

This is where the purpose of eco-design, or *oekodesign*[10], takes root, for its objective is to design a project, from sketch or feasibility phases (within the meaning of Act no. 85-704 of July 12, 1985, on public contracting and its relationship with private contracting, known as the MOP Act), according to ecological performance or co-benefit objectives. The aim is not to "wipe the slate clean" for the past but, based on technical engineering knowledge, to introduce biophysical considerations, in connection with the need to protect and develop the natural environment in the project to develop the sea.

1.2. Definition of eco-design of marine infrastructures

The eco-design of marine infrastructures is the result of recent cultural evolutions, as mentioned above. It is part of the interdisciplinary field of ecological engineering, which includes human sciences (geo-planning, law), engineering sciences (civil engineering, materials science) and natural sciences (biology, ecology). It responds to a major challenge for responsible human societies and biodiversity managers.

For Francis and Lorimer (2011), reconciling human and non-human in the project of developing urban territories, by integrating the conservation of nature, is undoubtedly the greatest challenge of the 21st century. For these same authors, the

---

8 *Center d'études et d'expertise sur les risques, l'environnement, la mobilité et l'aménagement* (previously CETMEF, before 2014).

9 See Chapter 4 of this book on the GIREL project at the port of Fos-sur-Mer.

10 We prefer this term to the one used in most English-language eco-engineering publications due to its proximity to the approach of project management for the design of a development project, even though they both designate the same field of reflection.

contribution of ecological engineering, a discipline that is still in its infancy in terms of the solutions it proposes, is a decisive factor in ecodevelopment. It was Odum, in 1962, in the context of his work on energy flows in marine ecosystems, who proposed the term "ecological engineering" (Odum *et al.* 1963). The most widely accepted definition today is that of B. Mitsch, a student of Odum, associated with S.E. Jørgensen (Mitsch and Jørgensen 1989): "Ecological engineering uses ecology and engineering to predict, design, construct or restore, and manage ecosystems that integrate human society with its natural environment for the benefit of both."

The intimate link between ecology, land use planning and civil engineering is underlined by Van Bohemen (2004) who makes it the key to its diffusion and application. According to Bergen *et al.* (2001), the design of development projects must integrate the issues of human societies in an ecological approach, for a mutual human–nature benefit, following steps that ratify its application, the first three of which are:

1) design in accordance with ecological principles;

2) design adapted to the environmental specificity of each site;

3) maintaining the functional requirements of the structures, regardless of environmental requirements.

While ecological engineering includes eco-design in its definition, this term seems important to us because it underlines a crucial aspect in the success of a project benefiting human and nature; that is, to apply ecological concerns at the feasability step, to guide the project design from the beginning. Indeed, all too often, so-called eco-designed infrastructures take into account environmental aspects once the technical-economic aspects have been resolved (Figure 1.2).

According to the French engineers' union Syntec Ingénierie, this approach can be defined as follows: "Eco-design is the technical design of projects by also considering global and local ecological concerns". It is "an approach that can be applied to a large number of sectors, without generating additional costs in the long run" (Les Cahiers de l'Ingénierie 2010[11]).

We will complete this definition with the idea that ecological added value can also improve the technical performance of a structure, as a co-benefit. This would be the case, for example, of a desired effect of bioprotection of an eco-designed quay enabling, by its colonization, the promotion of protection of concrete from chloride ions (gain on the aging of the structure which is avoided).

11 http://www.syntec-ingenierie.fr/actualites/2010/09/01/les-cahiers-de-lingenierie-n80-septembre-2010/.

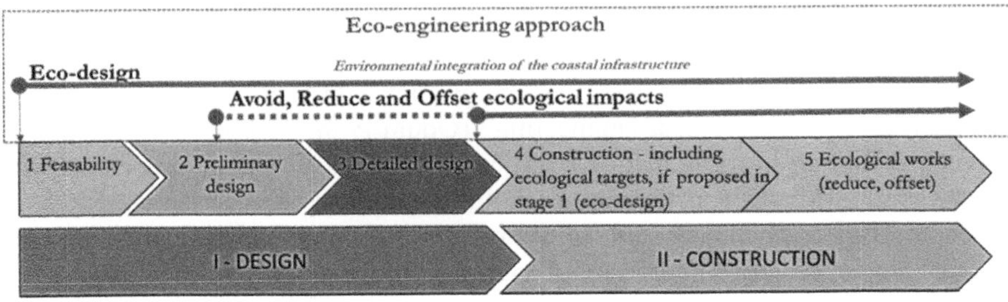

Figure 1.2. *Eco-design project methodology (top blue arrow) compared to a standard approach (middle slice of the figure)*[12]. *For a color version of this figure, see www.iste.co.uk/pioch/marine.zip*

We will therefore propose as a definition of eco-design of marine infrastructures the process of "designing sustainable maritime development projects with precise/specific technical and ecological functions[13], which generate socio-ecosystemic co-benefits, without generating additional costs in the long term".

It also seems important to us to address an aesthetic aspect for better integration into the underwater landscape. The aesthetic question of an object placed within a still "natural" environment, the seabed, seems essential to us. Indeed, addressing the "classic" needs of development meets precise rules of art, currently the subject of specifications, never integrating values of landscape aesthetics. However, the "beautiful" and "good" are always at the center of the concerns of humans, when they have the ambition to see their creations endure in the objective of being "well-made" and when they have the financial means to fully undertake this ambition. The morphologies of submerged elements are resolutely human and far from the soft forms usually found under the sea: cubes, trapezoids, beams, chaotic heaps, etc. A landscape approach would allow a better visual integration of these developments under the sea and undoubtedly a better acceptance of projects by the concerned actors and the general public. The contributions of bio-inspiration, more modest than the idea of biomimicry, can also usefully be called upon here, not only in terms of mechanisms that favor biodiversity (roughness, quality of materials, etc.), but also in terms of the overall visual aspect of a structure in a natural underwater environment. An eco-designed structure is a structure whose forms are bio-inspired and which

12 The eco-design approach begins in the early stage of the project, in comparison with usual eco-engineering treatments which are based on mitigation resulting from environmental impact assessments or ecosystem damage. They are mainly considered after the design of the project to reduce or offset significant negative impacts.

13 Specific, therefore assessable, quantifiable, required; specific, therefore adapted to each site or type of project requested.

integrates ecosystem conservation objectives into its functions, as well as its technical functions.

PIANC also emphasizes the need to design works according to the concept of sustainable development and publishes professional recommendations with this in mind (PIANC 2011a). This is the idea of "working with nature", developed by Dorien Korbee, which is defined as:

> an integrated process which involves working to identify and exploit win-win solutions which respect nature and are acceptable to both project proponents and environmental stakeholders. It is a philosophy which needs to be applied early in a project when flexibility is still possible (PIANC 2011b).

Nature-based solutions, popularized by the International Union for Conservation of Nature (IUCN), include infrastructure eco-design in their definition (Cohen-Shacham *et al.* 2016): "Nature-based solutions are actions built on ecosystems to address global challenges such as climate change or natural risk management", for example, the biogenic seawall for coastline protection (Figure 1.3).

Figure 1.3. *Diagram of the definition of nature-based solutions, proposed by the IUCN, where "conventional" infrastructure is cited as an application (Cohen-Shacham* et al. *2016). For a color version of this figure, see www.iste.co.uk/pioch/marine.zip*

Internationally, our definition of eco-design is related to the terms "green engineering", "eco-engineering", "nature-based solution" and "eco-design" (Pioch *et al.* 2018). It now seems interesting to us to better draw the spirit of this approach from the sources of inspiration of eco-design dedicated to marine environments, through the Japanese experience.

1.3. Japanese inspiration

It is in Japan, an archipelago country turned by necessity towards the sea, where the observation of nature is an art as much as a deep aspiration, that the ideas on the development of the seabed have been developed.

It was around the 17th century, during the reign of Emperor Jôo, that fish houses, reefs and artificial fishing habitats were built near the coast. It is within this culture of *Sato-umi*, literally "the sea where people live", and this vision of a fertile and rich sea for those who know how to change it that the Japanese idea of eco-design was developed (Yanagi 2012).

The very root of the word *œkodesign* (*oikos*, from the Greek word for home or house) is completely in line with that of *Sato-umi*, based around the notions conveyed by ecumene[14] and ecology. We can also find in this vision the inspiration developed by Rosenzweig and Michael (2003) for the ecology of reconciliation, where biodiversity and human development are closely intertwined in our "common home".

1.3.1. *Influence of the Japanese vision for sea-friendly development projects*

François Doumenge[15], in his book *Le Japon et l'exploitation de la mer* (1961), was impressed by "the incredible creativity of this people towards maritime productive development". He emphasized the empirical and traditional approach of Asian societies, where the observation of nature, expressed in the Shinto spiritual tradition (or *Kani-michi*, the "voice of the gods"), gave a familiar closeness between humans and a deified nurturing nature, but in proximity, not in superiority (Doumenge 1961, 1966). In fact, the spirituality developed around nature and Shinto is more a way of life than a religion in the Western sense, according to Doumenge. This vision has also influenced Augustin Berque (1996). He admits in his reflections

14 The relationship of humanity with the Earth's expanse.

15 Of the Languedoc Geographical Society.

to having been inspired by the Japanese author Watsuji Tetsurō to clarify his vision of the ecumene where the human is intimately dependent on the natural environment. The sea (*umi*) is both a vital and respected environment because it is a place of intense spiritual activity where the *kani* (familiar deities) live and where the souls of the dead spend time. The Japanese view of the sea, and the underwater world, is also one that is comparable to our view of our earthly, poetic and scenic landscapes (Figure 1.4).

Moreover, the two basic principles of Japanese coastal maritime law (Amako 2015) are based on:

– the "Yoro Code" (養老律令 enacted in 757): "coastal areas are of common (public) use and are managed by the users themselves";

– the "Ritsuryo Rules" (律令要約 enacted in 1741): "coastal areas are managed by delegation to local fishing communities".

The management of natural marine environments closest to the activities carried out there, in this case fishing, is a marker structuring Japanese thinking.

Figure 1.4. *Kushimoto Marine Park, protected for the value of underwater landscapes (photo: S. Pioch)*

1.3.2. *The Japanese* Sato-umi

This particular view, and the predominance of the sea, is explained by the geography of Japan, made up of five large islands and 6,847 islands and islets. The exclusive economic zone (EEZ) of Japan, with 4,050,000 km², is also the sixth largest in the world.

With only 14% of the country's land area available as arable land, the submerged volume of the sea has long been an essential element in providing the country's food resources, along with rice cultivation. The ICZM vision is therefore naturally developed in Japan under the term *Sato-umi* (Figure 1.5).

The term was proposed by Yanagi; it comes from its terrestrial counterpart *Sato-yama*, where *sato* refers to a place where people live and *yama* to a forest; the combination of the two terms then leads to a third definition: "the forest near the place where people live". *Sato-yama* came from a reaction that we might call neo-modernist, questioning the idea that nature would reach its fullest potential without the work of humans (Dublin *et al.* 2014). This is what is developed by Lassus (2002) in his book *Harmonie et règles urbaines*, where the work of humans respects biodiversity, creating different interactive biotopes.

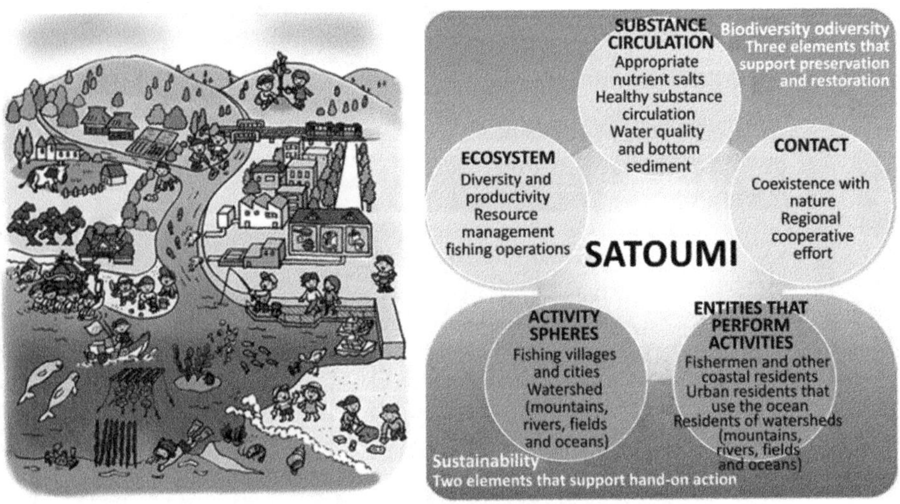

a) b)

Figure 1.5. *The concept of* Sato-umi, *where humans interact with the aquatic environment for mutual benefit: (a) Japanese Ministry of the Environment (2015); (b) Yanagi (2007)*

This idea of *Sato-umi*, close to the European vision of ICZM[16], reinvests the contributions of ecological engineering to integrate nature into the design of infrastructures.

1.3.3. *Artificial marine habitats: Japanese origin of inspiration for eco-design*

Japan is undoubtedly considered as the world leader in the successful management of the seabed, as much in terms of application (stability of coastal fisheries) as perspective (Bailly 1989; Baine 2001). Extensive coastal mariculture (or marine-ranching) began with the great interest of this maritime nation in the creation of "fish houses", habitats adapted to the needs and functions necessary to increase underwater life. Thus, the first Japanese reefs, or artificial habitats (AHs), were described between 1652 and 1655 (during the reign of Emperor Jôo). They consisted of wrecks filled with stones to extend the exploitable fishing areas near the village. The technical description is very precise, and the choice of materials (stones and timber from ships) is linked to questions of habitat (volume, ecological functions, target species, stage of development) and durability. However, it was not until the 20th century that this structural diversity really developed in Japan, and then from the 1950s onwards for the rest of the world.

The first modern (disseminated) Japanese studies on AHs date back to 1903 (Fukuoka Prefecture), with the most important trials starting in 1952 with the first governmental plan for "development of shallow water fishing areas" (Simard 1996). The Japanese public investment for the creation of AH infrastructures in 2007 alone was 602 million euros (93.318 billion yen), with an average between 2000 and 2005 that was fairly stable at approximately 600 million euros per year (Japanese Institute of Fisheries Infrastructure and Community[17]). In 2001, an audit launched by the French Ministry of Agriculture, Fisheries and Forestry (MAFF) found that more than 12% of the continental shelf was developed with about 20,000 sites (Lacroix *et al.* 2002). In 2007, there were over 222 AH models referenced by the JIFIC (Pioch 2008). These figures allow us to imagine the immense amount of technical and ecological knowledge that has been developed to understand the marine species/habitat relationship. We will return to the contribution of this work, which we use to design structures that deploy habitats and functions adapted to local species, in Chapter 3.

NOTE.– In Europe, habitats are defined according to the European Nature Information System (EUNIS) habitat (and settlement) classification method. This typology of habitats is the most commonly used. It is a habitat classification

16 Integrated coastal zone management.

17 JIFIC: https://www.maff.go.jp/e/.

reference for terrestrial, freshwater and marine areas (defined via the Rebent Ifremer network). This typology has been used to determine the major physical habitat types of the French seabed. It is also used to characterize the determining habitats of marine natural zones of ecological interest, fauna and flora (ZNIEFF). This typology is based on a hierarchical classification of habitats, which can be measured to different levels of precision, ranging from the simple distinction between rocky and soft types (level 2) to the precise identification of benthic populations (levels 5 and 6). These benthic communities are characterized at the level of each major habitat type: a major benthic habitat type corresponds to one or more habitat types as defined in the EUNIS classification (Bajjouk *et al.* 2015).

This national reference system defines a common language to facilitate the implementation of public conservation policies and makes it possible to inventory the habitats present in metropolitan France. Its use is recommended for all programs related to the conservation of marine habitats on a national scale.

New regulatory requirements are leading to changes in coastal development practices. The purpose of this work is not to recall the regulatory principles relating to environmental or urban planning law; we refer the reader to the many specialized works on the subject. However, we believe it is useful to indicate recent developments, as they corroborate the interest of project owners and their consulting firms in applying an eco-planning approach.

Indeed, protecting the environment is not only about conserving spaces and species and classifying territories to protect them from human activities. It also means integrating the environment into planning actions and development projects, which are the subject of this book, and therefore designing projects that are respectful of humans, landscapes and natural environments and concerned with saving space, sparing species and limiting water, air and soil pollution.

1.4. Assessing the effects of a project on the environment: the impact study

As we briefly recalled at the beginning of this book, humanity's priority is to deal with the artificialization of natural environments, which is dramatic for biodiversity, particularly in marine and coastal environments. Among the tools available to planners who are aware of this global problem, the environmental impact assessment (EIA) of projects has been widely used since the 1970s (Bonneuil 2004). This tool is inspired by work in ecology intended to organize the comparison of the spatio-temporal effects of a human action on a natural environment (before/after an impact), formalized by the Before-After Control-Impact (BACI) approach (Green 1979).

The EIA is thought of as a regulation of unreasonable artificialization and as a tool for social acceptability; it is integrated by the state, planners and civil society (Gobert 2010). In its most modern form, particularly in France, it is based on three hierarchical sequences for mitigation in the analysis of the environmental impacts of a project, symbolized by the acronym ARO for "avoid, reduce, offset", and applied in a manner proportionate to the issues at stake in a project.

The pivotal consideration of the impact assessment, whether or not it leads to development authorization, and thus to construction choices, clearly rests on this sequence, where offset is supposed to negate the notable and residual negative impacts of a project (Pioch *et al.* 2015).

Assessing impacts is not straightforward. This is what road ecology deals with, for example, in an attempt to quantify the impacts (positive or negative) of linear infrastructures on the environment in general, and more specifically on biodiversity, with effects of habitat destruction or fragmentation (Clevenot *et al.* 2017). Part of the research is devoted to mitigating and offsetting for negative impacts. The aim is to make infrastructures "transparent" to the landscape and its ecological function (Rytwinski *et al.* 2016).

Thus, over the past 50 years, planners have had to develop new approaches to meet this requirement to avoid, reduce as much as possible (or at least below a "significant" threshold) and finally offset the inevitable or irreducible residual impacts of any project. However, even though the regulations seem to be robust, if we look at the results in France, let us be clear, the record has been catastrophic since the introduction of the EIA in 1976 (Act no. 76-629 of July 10, 1976). According to Bigard *et al.* (2017), more than 77% of the definitions of ARO measures proposed in a sample of 42 terrestrial EIA files are wrong. At sea, Jacob *et al.* (2016) demonstrated, based on the study of 55 EIA files between 2000 and 2015 (France and DOM-TOM[18]), that 85% conclude that there is no significant residual negative impact of infrastructures on biodiversity. Furthermore, in the 15% of projects identifying the need for compensatory measures to repair biophysical losses, these are not actually implemented in the field (less than 10% of cases). These data and observations lead us to take a critical look at our own development (or "non-development") choices.

The fundamental objectives of the EIA are recalled in Article L. 122-1 of the Environmental Code, requiring that "the studies prior to the realization of developments or works which, through the importance of their dimensions or their

18 *Départements d'outre mer* and *Territoire d'outre mer* (French Overseas Departments and Territories).

impact on the natural environment, may affect the latter, must include an impact study to assess the consequences".

The impact study is an essential step in the environmental assessment of development projects and works. It is an approach designed to:

– integrate environmental concerns into the design of a project by its promoter;

– inform the authority called upon to decide whether to authorize its implementation;

– inform the public by involving them in the decision-making process.

1.4.1. *Designing a better project for the environment*

The impact study is the means for the project promoter to verify that environmental concerns have been respected. As such, it must be an effective tool for integrating the environment into development projects and thus participate in the design of projects that respect people, landscapes and natural environments and are concerned with saving space, sparing species and limiting water, air and soil pollution.

1.4.2. *Informing the administrative authority on the decision to be taken*

The impact study is a prerequisite for an administrative decision to authorize or approve a facility, structure or development. It contributes to:

– informing the competent authority, that is, the administrative authority (minister, prefect, mayor, president of a general council) that authorizes the work, about the nature and content of the decision to be made (authorization, approval or refusal of the project);

– guiding the above in defining the conditions under which this authorization is given, for example, with regard to the implementation of measures to eliminate, reduce and offset for harmful effects;

– controlling *a posteriori* the respect of the commitments made by the project owner, for example, by ensuring a follow-up of the consequences of the project on the environment during the implementation and operation phases.

In France, the procedure required by the legislator is broken down into 11 sections described in Article R 122-5 of the French Environmental Code. For the study of the project's impact on the environment, these 11 points can be grouped into six stages, as described in Figure 1.6.

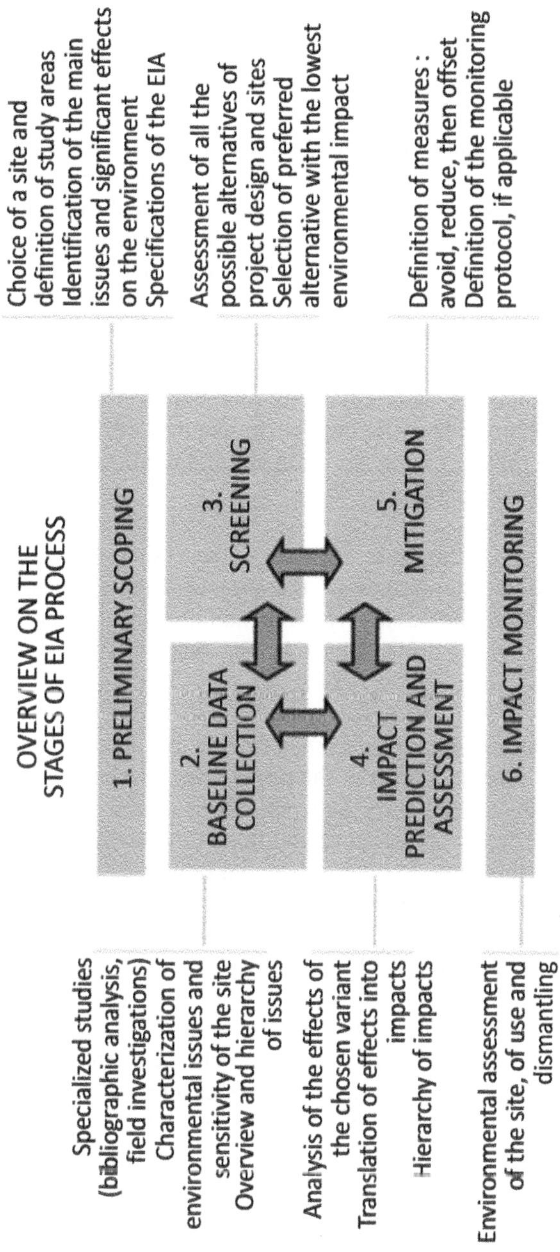

Figure 1.6. *General approach to conducting an environmental impact assessment (French Ministry of the Environment)*

1) Preliminary scoping (step 1/6): this preliminary phase after the project proposal and screening, carried out by the project owner, with or without technical assistance (AMO), consists of identifying a limited number of potential environmental issues from the vast field of those to be dealt with in depth in the EIA. The project owner can officially request a scoping exercise from the investigating department to help identify the priority issues and other projects to be taken into account in the analysis of cumulative impacts, as well as the data sets or studies available on the project site. This framework makes it possible to determine whether the project should be subject to a declaration or an application for authorization under the Water Act, according to the headings of the nomenclature resulting from Article R. 214-1 of the Environmental Code which may apply to it.

2) Baseline data collection (step 2/6): the objective in this step is to provide *ex ante* the environmental status of a site. This baseline is based on data from the biological, physical, landscape and human environment of the site before the project is implemented. It must provide sufficient data to identify, evaluate and prioritize the possible effects of the project on the environment and to allow for the rehabilitation of the site at the end of the operation.

3) Screening (step 3/6): this step consists of assessing, through multi-criteria analysis, all the possible alternatives of the project design and sites to select a preferred alternative with the least environmental impact, excluding alternatives with an unacceptable impact. It also prepares for addressing the significant impact with steps 4, 5 and 6.

4) Impact prediction and assessment (step 4/6): at the intersection of the initial state and the technical characteristics of the project, the EIA includes "an evaluation of the direct and indirect, temporary and permanent effects of the project on the environment, and in particular on the fauna and flora, the sites and landscapes, the soil, water, air, the climate, the natural environment and biological balance, the protection of property and cultural heritage and, where appropriate, the convenience of the neighborhood (noise, vibrations, odors, light emissions) or on hygiene, health, safety and public health" (Article R. 122-3 of the French Environmental Code).

5) Mitigation (step 5/6): mitigation measures aim to avoid, reduce and finally offset some of the impacts envisaged despite the choice of the alternative with the least environmental impact (priority avoidance measures). This ARO sequence is essential and must be treated with particular care. Indeed, if rehabilitation measures can be envisaged at the end of the EIA, it is always better controlled and often easier to avoid or limit degradation rather than offset it. This phase of the EIA also involves identifying monitoring protocols (methods, indicators, plans and sampling strategies) that are most relevant and appropriate to the effects being monitored. This part is usually the core of the environmental impact statement used for the EIA review by the competent authority (regulators) and other consultees, including the public and NGOs.

6) Impact monitoring (step 6/6): finally there is the impact monitoring step. It consists of an EIA follow-up by determining the protocols for monitoring the effects of the project and the mitigation measures over a variable period of time depending on the project, but including minimal monitoring before the start of the work (analysis that can be carried out during the initial state), monitoring during the construction phase, monitoring during the exploitation phase and, in the case of temporary occupation of the public maritime domain (PMD), monitoring following the dismantling of the exploitation unit (offshore wind farm, underwater pipeline, etc.). If necessary, monitoring also focuses on the effectiveness of offset measures (specific monitoring).

It should be noted that the preparation of an EIA is an iterative process and these six hierarchical sequences do not constitute successive, compartmentalized stages. In order for the EIA to serve as an effective decision-making aid for the project developer and its stakeholders, it is necessary to return to the different sequences several times, in iterations. For example, successful design of mitigation measures could possibly result in the removal of all significant impacts; hence a new screening exercise would reveal that there might have been no need to carry out a formal EIA had the mitigation measures been included from the start.

Once again, the EIA must be initiated as early as possible and follow a continuous, progressive and iterative process based on a good knowledge of environmental issues.

IMPORTANT.– *Always give priority to avoidance; as mentioned above, avoidance is the only solution that will ensure that the project does not damage the environment and that major issues are preserved.*

1.5. The "no net loss of biodiversity" objective: a regulatory obligation for developers, an opportunity for eco-design

According to the Environmental Code, the application for authorization of a project likely to affect the environment must be preceded by an assessment of its environmental consequences. The content of the impact study has just been extensively reformed by a decree and ordinance that reinforce the respect of the hierarchical "avoid, reduce, offset" sequence (ordinance no. 2016-1058 of August 3, 2016, decree no. 2016-1110 of August 11, 2016).

The Act of August 8, 2016 on the reconquest of biodiversity adds to and crowns this normative reinforcement by enshrining the concept of "no net loss of biodiversity" in French law.

From now on, any development project must produce a net neutral balance[19] with respect to the losses it will cause to biodiversity in terms of species, habitats, functions and finally ecosystem services.

The impact study is both an approach and a regulatory file:

– the approach is a thorough reflection on the impact of a project on the environment, conducted by the project owner in the same way as they study the technical and economic feasibility of their project;

– the file is the document that explains, in particular for the attention of the examining body that issues the authorization and for the attention of the public, the way in which the project owner has taken the environment into account throughout the design of their project and the measures to which they commit themself to mitigate its impacts.

Regulatory reinforcement now requires a new vision of the development project, not only socio-technical-economic but also ecological, in order to minimize (cancel?) the negative impacts.

These requirements are constantly being reinforced. Even though the controls on the ground are still too weak, it is undeniable that the application of the regulatory framework aiming at a "no net loss of biodiversity" will reinforce the interest towards constructions with a reduced effect on the environment, or even positive ones.

However, a development project exists in response to a functional need that requires a technical response: (i) the function of a breakwater is to protect a port and is adapted to the hydrodynamic actions of the place; (ii) a quay is there to allow the berthing, mooring and loading/unloading of ships and is adapted to the project ship and the transported goods.

This idea should not replace a deleterious business as usual from human activities damaging to the environment, thanks to green washing (fake eco-design) that allows future construction. It is very important to underline here that this approach is to be carried out in parallel with the ARO approach in the impact study.

Therefore, before any integration of the environment into a technical design, it is absolutely necessary to perfectly master the latter. A certain technical virtuosity is

19 Goal of "no net loss" of biodiversity, Act No. 2016-1087 RBNP of August 8, 2016.

required to propose daring technical variants necessary to take the environment into account in the eco-design process.

This is the objective of Chapter 2, which will present a brief review of the technical concepts related to marine structures and the skills that the marine engineer must have.

1.6. Specificities of the environmental assessment related to the marine environment: the "natural" public maritime domain concept

1.6.1. *The natural public maritime domain, an essential concept*

An important specificity of an eco-design approach in the marine environment is the legal framework attached to it as public property. The natural public maritime domain (nPMD)[20] refers to this remarkable, sensitive and coveted space at the interface of land and sea (Figure 1.7).

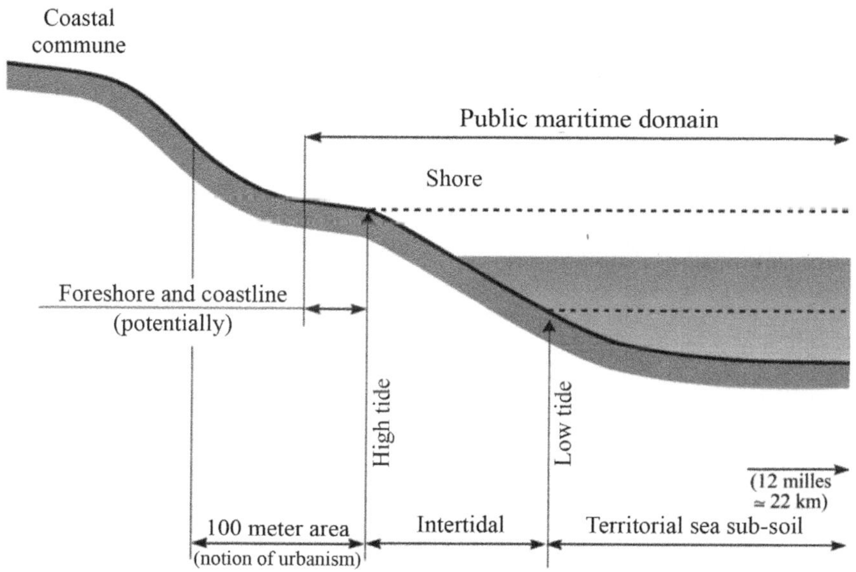

Figure 1.7. *The extent of the natural public maritime domain in France (Article L. 2111-4 of the CGPPP). For a color version of this figure, see www.iste.co.uk/pioch/marine.zip*

20 The regulations relating to the nPMD were recalled and detailed by the circular of January 20, 2012, relating to the sustainable and integrated management of the natural public maritime domain.

In France, reflections on its status as a "natural" space are long-standing, dating back to the Edict of Moulins (1566) and to Colbert's ordinances on the Navy (article 1 of title VII of book IV of the ordinance of August 1681): "The edge and shore of the sea will be considered to be all that it covers and uncovers during the new and full moons, and as far as the great flood of March can extend over the shores."

This "natural" territorial zoning has been confirmed by jurisprudence, in particular the French Council of State decision which specifies "the limit of the nPMD at the point of the highest seas [...] in the absence of exceptional disturbances" (Kreitmann decision of October 12, 1973).

The nPMD cannot be the object of private appropriation. The marine environment is therefore considered as *res nullius*, belonging to no one, without master or owner, but which can be appropriated in the context of uses (fishing, exploitation of aggregates, energy production, etc.), and its resources *res communis*, belonging to all citizens, therefore accessible and usable by all.

At the global level, this principle, and the resulting jurisdictional divisions, are a constant in all coastal states (Féral 2011). The roots of this conception of the marine environment can be found in the marine code of Eleanor of Aquitaine (12th century), followed by the claim of the Dutchman Hugo Grotius in 1609, for a *mare liberum* linked to the objective of free trade at sea. Only the coasts were subject to a special provision for the allocation of their resources and the protection of state borders.

The sovereign power of a coastal state extended pragmatically to what could be defended, that is, 3 nautical miles (1 nm = 1.8 km), or roughly the range of a cannon shot, from the coast. The 1982 Montego Bay Convention (UNCLOS) modified this conception of a sea as a "public asset".

With the exception of this narrow strip of 3 nautical miles, the rest of the maritime continent was a "free" or lawless zone until the last 30 years. From then on, the "territorial seas" were extended to 12 nm (about 22 km) from the coast, with a contiguous zone of 22 nm where only the right of prosecution of states is exercised, within the framework of customs operations (drugs, human trafficking, etc.). The UNCLOS developed the idea of exclusive economic zones (EEZs[21]) that reach up to 200 nm from the coastline (Figure 1.8). In the EEZs, states do not exercise formal sovereignty but have the sole right to exploit resources (fisheries and fossils) from the surface of the ocean to the seabed and its subsoil.

21 United Nations Convention on the Law of the Sea, signed on December 10, 1982.

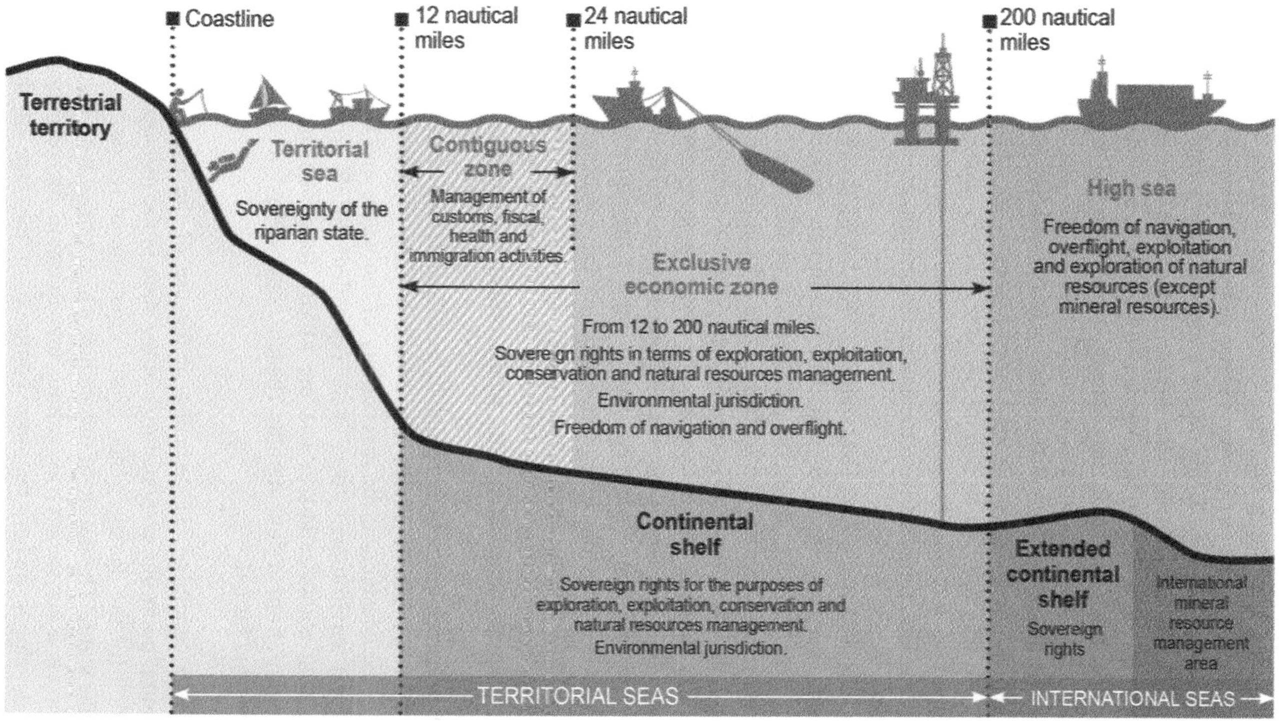

Figure 1.8. *Maritime boundaries and spaces of coastal state jurisdiction (MEB 2016). For a color version of this figure, see www.iste.co.uk/pioch/marine.zip*

In addition, states may request the extension of their EEZ to a maximum distance of 350 nautical miles, depending on the geomorphology of their continental shelf (geological continuity extension). However, the principle of free access, of *mare liberum*, always prevails beyond EEZs on the high seas. It is a free commodity that belongs to no one, does not fall under the sovereignty of any state and is not subject to any appropriation regime, nor to governance (*res nullius*). The threats to fish stocks are therefore still very real. In contrast, it is within the framework of the UNCLOS that a new notable status was established for the sea, that of a World Heritage Site. The founding act of this idea was the speech made in 1967 by Arvid Pardo, Maltese Ambassador to the United Nations (UN 1967, 22nd session, meeting 1515). He was the first to defend the idea that the seabed and its subsoil should be defined as a world heritage site: a space of goods belonging to all humanity and, therefore, removed from the exclusive appropriation of states.

The principle was confirmed by Declaration 2749 (XXV), adopted by the United Nations General Assembly in 1970, and then definitively endorsed in the 1982 Montego Bay International Convention on the Law of the Sea, which created a special legal regime applicable to the seabed and its subsoil beyond the EEZ. In this part of the sea, called the "zone" (Figure 1.8), states or industrialists are allowed to explore or exploit mineral resources upon approval by the International Seabed Authority. However, at the same time, the mechanism proposed that an organization called "The Enterprise" (never defined) would exploit the same quantities as the industrialists (another principle of equivalence), but for the benefit of developing countries. A beautiful vision, but which in fact has not yet given rise to any "repayment" of the exploitation of this immense area, for the development aid of the planet, by an exploiting industrialist. Technical problems or timid philanthropism?

It is indeed this general principle of a public "asset", under the sea, which influences the development of its coastline. It also establishes a binding obligation, applied to the right to develop the coastal sea: the reversibility of the works and the return of the developed natural environment to a state equivalent to its initial state. The concessions or self-authorization to use the nPMD (environment) are therefore temporary, ranging from a few months (beach concessions) to several decades (ports, offshore wind farms).

This "maritime condition", in particular of the use of the marine environment, introduces a central point: the responsibility of the state in the management and thus the compensation for the damage made to this environment (or domain, due to its attachment to the state). The notion of ownership, which carries with it an obligation of management vis-à-vis society, by calling into question the responsibility of a "polluter" or developer owner, becomes confused in the face of the size and sovereign position of a state that certainly manages, but which "self-controls" (Pioch

2008). Moreover, in a context of decreasing state resources, how can we guarantee effective control of the implementation of compensatory measures? How can we monitor them? Should these legitimate objectives be transferred to marine environment management organizations that have already been delegated by the state, and, if so, how can we succeed, within a constant financial perimeter, in requesting new missions of the field agents of the French Office for Biodiversity, the *office de la biodiversité* (formerly the *agence des aires marines protégées*, the French Marine Protected Areas Agency) or the *conservatoire des espaces littoraux et des rivages lacustres* (CELRL, the French Coastline and Shoreline Conservation Authority)?

Added to this is the difficulty inherent in monitoring the "underwater world", where the human is a temporary guest and where the degradation of an underwater landscape (still a vaguely defined concept) is not so familiar.

1.6.2. *Managing the marine environment: a necessarily integrated approach*

At the maritime level, the Marine Strategy Framework Directive (2008/56/EC), adopted in 2008, is a European directive that establishes a framework for community action in the field of marine environmental policy.

This directive is based on a so-called "ecosystem approach": it aims to achieve a good ecological status for the marine environment by 2020 and to improve the conservation status of marine biodiversity from 2015. The main motivation of the directive is to combat the many threats to the marine environment, such as:

– the depletion or degradation of biological diversity and changes in its structure;

– habitat loss;

– contamination by hazardous substances and nutrients;

– the impacts of climate change.

These points require a global framework to coordinate the local actions of member states.

In contrast, as early as 1975, the Council of Europe recommended a global treatment of the coastline. A European coastal charter was signed by the Conference of Peripheral Maritime Regions in 1982. Chapter 17 of Agenda 21, adopted at the Rio Conference in 1992, recommends ICZM, linked to the protection of the marine environment. An ICZM approach aims to enable a global approach to a coastal territory by taking into account the physical context (geomorphology, weather, tides, marine currents, etc.), the ecological and biogeographical context, the socio-economic context and the legal-administrative context. The integration sought concerns, in particular, the sectoral (interactions between actors), environmental (ecosystem approach), geographical and biogeographical dimensions, as well as the integration of governance (participation of all the actors concerned in the decision), which is an essential dimension. The final objective of ICZM is to build structures and regulatory instruments that will make it possible to guarantee or restore the balance between human activities and human and natural resources, in order to avoid overexploiting resources that are not slightly, easily, slowly or expensively renewable, especially fisheries and tourism resources, with the general aim of ensuring the sustainability of development.

In France, the Coastal Act, adopted in 1986 (Act no. 86-2 of January 3, 1986 relating to the development, protection and enhancement of the coastline), constitutes the reference text for the organization of sustainable development of coastal areas. Let us briefly recall its main objectives, which are:

– the protection of biological and ecological balances, the preservation of sites, landscapes and the cultural and natural heritage of the coastline;

– the preservation and development of economic activities related to the proximity of water;

– the implementation of a research and innovation effort on the particularities and resources of the coastline.

This act applies to municipalities bordering seas and oceans, salt ponds and inland water bodies with an area of more than 1,000 hectares. The municipalities bordering the PMD are considered to be those bordering the sea and the ocean. Consequently, in estuaries, only the municipalities located downstream from the transverse limit of the sea (the limit between the PMD and the public river domain) belong to this category of "coastal municipalities" by right.

Outside of the already urbanized areas, all construction is prohibited in a strip of 100 m from the high limit of the shoreline. This 100 m strip constitutes a minimum which can, through the application of Article L. 146-4-III C. Urb., be extended by the local urban plan, when the sensitivity to the environment or the erosion of the coasts justifies it. This prohibition does not apply to installations requiring the immediate proximity of water. Beyond that, in the areas close to the shore, a regime of limited extension of urban development is provided for. Notable and characteristic areas of the coastal municipalities are preserved by a strict regime of inconstructibility. Implementation of these principles must be completed by the plans for the development of the sea.

Among the normal uses of the PMD between which the administration has to arbitrate, we can mention:

– the implementation of port and maritime safety works;

– the use of beaches for bathing;

– the reception of marine cultures on the favorable spaces;

– retention as a natural area.

2

Maritime Civil Engineering

In this chapter, the authors return to the fundamentals of maritime civil engineering. Indeed, the eco-design approach requires the engineer to have advanced technical skills in order to be able to respond to the client's functional problematic, propose simple, clever and daring solutions that guarantee an optimal technical design, respect the economic and societal constraints of the development and propose a successful integration of the development into its environment.

Achieving these objectives is difficult, and to do so it is necessary to review the entire technical theme: (i) the general concepts relating to marine developments and their integration with the coastline; (ii) the constraints of the project and the knowledge of the physical environment; (iii) the expected characteristics of the works in response to the requested functionalities.

In order to address these different themes, we have drawn on our experiences developed during our careers as engineers and teacher-researchers.

In this chapter, we thought it would be useful to cross-reference our experiences with other maritime experts, namely Dr. Jean Bougis, consulting engineer[1]; Jean-Michel Cathala, former head of the civil engineering department of BCEOM[2]; and Bernard Sigros, former head of the design office of EMCC[3].

Through this approach, we also wish to highlight the technical and human greatness of these three high-ranking engineers, who have accumulated more than

1 https://scs-ingenierie.com.

2 A French engineering, design and project management company, now EGIS Eau.

3 The company has become VCMF (Vinci Construction Maritime et Fluviale).

120 years of experience between them. We thank them warmly for their contributions[4].

2.1. General information

The sea is an immense source of "wealth" for humanity: (i) as a vehicle for the transport of people and goods, from the start of humankind, it has connected the continents; (ii) as a source of living resources, it has provided major contributions in terms of food and the manufacture of medicines and other products useful to humankind; (iii) it provides mineral and energetic resources to humankind; (iv) it offers humankind a coastline which is a privileged living environment for us. The result is the development and settlement of humans on the coastal strip, which leads to strong land pressure on the coastline, both for human habitation and for industrial or leisure infrastructures. With an increasing intensity over the last 100 years, currently 60% of the world's population lives in the large coastal zone (~4 billion people within 150 km of the shore according to the IUCN). Artificialization is rampant, and it is estimated that 5–10 km of coastline containing nursery areas, coastal mudflats or mangroves are, if not polluted, deteriorated or destroyed (Béoutis *et al.* 2009). This global context of high coastal demographics is reflected in France where 885 coastal communities host a little over 10% of the population on only 4% of the territory!

This intense human coastal development therefore requires a need for service infrastructures comprising ports and coastal cities. These infrastructures are exposed to the sea, a physical environment capable of developing considerable destructive forces. It is, however, vital to maintain the coastline and it is therefore necessary to build coastal defenses, not to protect the coastline, exposed to the sea for the past millennia, but to safeguard the supposedly worthwhile human developments, established according to the social and environmental rules recalled in Chapter 1 (even if we have seen that, currently, the application of these rules through the impact study is extremely open to criticism, on land as well as at sea, and absolutely not reflective of the stakes involved).

The maritime engineer must therefore be able to combine solid technical knowledge with a great deal of modesty in the face of the natural elements and the resulting forces. They must be aware that a raging sea can deliver an incomparable force at our human resources. The first level of technical response of the marine

4 The combined experience of J.-M. Cathala, B. Sigros and J.-C. Souche is the subject of a maritime course for final-year students in the "Infrastructure and Major Works" program at IMT Mines Alès. The collaborations between J.-C. Souche and J. Bougis are recurrent and periodic in the context of expertise and consulting missions for the design of floating marine structures.

engineer is therefore the understanding of the coastline and the physical phenomena at play in the projected location of the development.

Second, for any maritime project, it is necessary to quantify the effects of the sea that will impact the planned infrastructures. However, while we can increasingly better predict the periodic servicing of ships and structures imposed by the sea, it also generates ultimate random actions, which are difficult to determine and are evolving under the effect of climate change. Only *stochastic* models, which transform the statistical knowledge of the past into probabilities for the future, allow us to understand these values, but with some uncertainty. However, they allow us to propose resilient constructions, provided that damage and ongoing maintenance are acknowledged. This is one of the notable features of maritime structures: they are designed for a period of use associated with a level of resistance and a durability requirement (Harshada *et al.* 2017). This difference from land-based structures is simply due to the aggressive (physical, chemical and biological) environment that the sea represents for construction.

However, this sea, whose power humans will never equal, remains a fragile ecosystem. Accidents, often resulting from mistakes made consciously or unconsciously by humans, can degrade the life that has been established there for billions of years, in a very damaging and lasting way. The countless oil spills around the planet are one of the most visible illustrations of this. Other impacts, such as the effect of tanning creams used by tourists, are much less visible. There is also the visible, but above all invisible, pollution from plastics, and their residues in the form of microplastics, with unknown consequences for all living beings, which are now being revealed after decades of massive "normal and usual" pollution.

Lastly, let us not talk about the cumulative effects of climate change from repetitive and increasingly significant anthropogenic impacts, which are leading to an unprecedented environmental degradation of the oceans (the sixth mass extinction of species), a phenomenon of which awareness remains recent, late, with almost no practical response.

With these facts in mind, the marine engineer can no longer design infrastructure with a negative environmental balance. As stated at the end of Chapter 1, the credo of "no net loss" is essential. Eco-design is one of the elements of response, admittedly modest in the face of the need for socio-cultural change, but important in the short and medium term.

However, before reaching this ambitious goal, it is necessary to detail the technical vision of the marine engineering project and the understanding of the environment, which starts with the observation and reading of the coastline and the understanding of the natural biophysical constraints that will be applied to the projected infrastructure.

2.2. Typology of coastlines[5]

2.2.1. *Coastal classification and sea-level evolution*

The coastline is the dividing line between the land and the sea. This limit is subject to two types of variations:

– variations on a geological scale, such as marine transgressions and regressions, erosion and sedimentation, etc.;

– daily variations due to tides and waves, to which should be added less systematic variations, such as storms and tidal waves.

The most recent classification was proposed in 1952 by F.P. Shepard. It distinguishes between primary coasts, which result from a contact between land and sea, shaped or not by a continental process, and secondary coasts, the morphology of which results from the current marine evolution.

Primary coasts are divided into four main groups: (i) coasts modeled by erosion (fjords, coasts with creeks); (ii) coasts resulting from the accumulation of continental deposits (deltas, alluvial plain coasts); (iii) coasts of volcanic origin (ring forms); and (iv) coasts of tectonic origin (drowned tectonic folds or faults).

Secondary coasts are, most often, former primary coasts remodeled by the marine environment. They are divided into two main groups: (i) coasts formed from erosion (denuded coasts or cliffs); and (ii) coasts resulting from accumulation (coasts with marine deposits such as sandy spits, beaches, lidos, tombolos, etc.) or even built by living organisms (corals, mangroves, etc.).

The littoral zone can be broken down into four stages, the relative importance of which depends on the type of littoral (Figure 2.1):

1) supralittoral, above the level of the high tide, but still subject to marine influence;

2) mediolittoral, within the tidal range (or intertidal);

---

5 This section is based on the work of J.-P. Pinot (1998) and the courses of J. Bougis. Source: https://scs-ingenierie.com.

3) sublittoral, between low tide and the beginning of the lower limit of the euphotic zone (about 30–50 m, the level of large algae and seagrass beds, varying according to the turbidity);

4) circalittoral, to the edge of the continental shelf.

The neritic province comprises the nearshore marine areas, up to the edge of the continental shelf (−200 m). Benthic species live on the bottom and sessile species, such as corals, algae and seagrass beds, are fixed (and are the most impacted by developments), while vagile species, such as lobster or scallops, move; as for pelagic species, such as tuna, they move in the water column.

The euphotic zone, as opposed to the aphotic zone, is a zone where light can penetrate and photosynthesis is still possible.

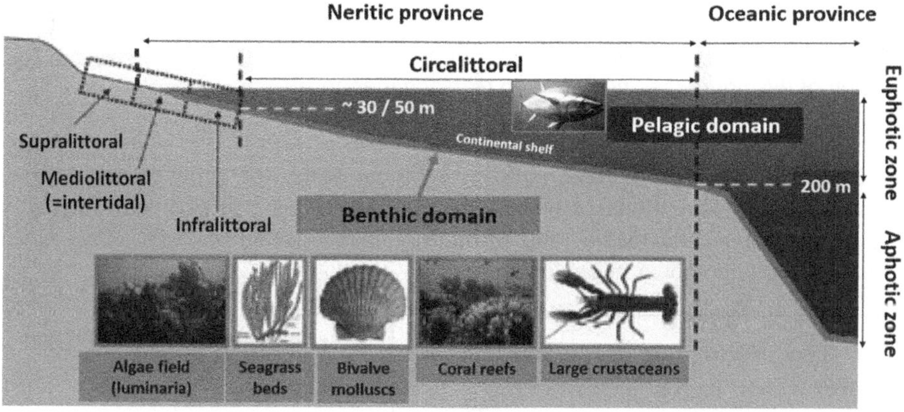

Figure 2.1. *Coastal stages, domains, zones and maritime provinces. For a color version of this figure, see www.iste.co.uk/pioch/marine.zip*

The existence of tides affects the distribution of living organisms, and it is considered that 80–90% of the marine living world, animal and plant, lives in the neritic province corresponding to the continental shelf (although many recent discoveries on ecological hotspots near deep hydrothermal springs tend to reduce this average). This zone represents barely 7.5% of the marine waters and 14% of the environments of our planet. It includes the three coastal submerged stages with 27,100,000 km² out of the 361,000,000 km² of the total surface area of the world's marine waters (see Figure 2.2), which themselves cover 70.78% of the surface of the earth.

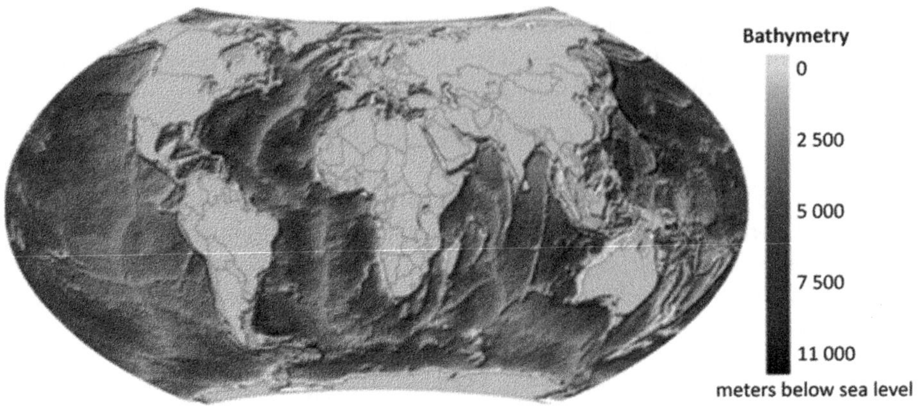

Figure 2.2. *World surface covered by the continental shelf represented in bluish white: medio-, sub- and circalittoral zones, submerged sections of the littoral stages up to the –200 m isobath (source: World Data Center for Marine Geology & Geophysics (2001); 2-Minute Gridded Global Relief Data (ETOPO2)). For a color version of this figure, see www.iste.co.uk/pioch/marine.zip*

Despite their small surface area on a global scale, the productivity of these environments is exceptional, especially in the areas that are of particular interest to our work, that is, the medio- and infralittoral zones (seagrass beds, coral zones, estuaries, etc.), as shown in Table 2.1.

Type of ecosystem	Area (10^6 km²)	Average net primary production (g/m²/year)
Continents	149	773
Ocean (offshore)	332	125
Continental shelf (-200 m)	27.1	360
Estuaries	1.4	1,500
Coral reefs and seagrass beds	0.6	2,500

Table 2.1. *Average net primary production and surface area of the major global ecosystems (Ramade 2009)*

Thus, the closer we get to the coasts, the greater the production of natural marine environments, but they are also, by proximity, the most exposed to anthropic activities and pressures. This productivity is accompanied by a significant diversity of life forms. If we were to divide metropolitan France into three large areas according to their sedimentary nature, the physical–chemical substratum and the quality of the water and the development of marine life, we could obtain:

– rocky seabeds, associated with Posidonia meadows (Corsica, Provence-Alpes-Côte d'Azur up to Port-de-Bouc and the Gulf of Lion from Palavas-les-Flots to the Cap d'Agde then from Argelès to the Spanish border);

– rocky seabeds, associated with algal meadows often dominated by laminaires (from Hendaye to Biarritz, then from the north of the Gironde to the Cotentin and a few scattered rocky points around Étretat or between Boulogne-sur-Mer and Calais (Cap Gris-Nez));

– sandy or pebbled seabeds (the Gulf of Lion, the coasts of Landes, Gironde and Normandy (Caen) up to the Belgian border).

Most of the French overseas departments and territories are located in the tropical coral zone, including the French West Indies, Polynesia, New Caledonia, Mayotte and Reunion, except for Saint-Pierre-et-Miquelon, whose seabed is similar to that of the rocky English Channel, and the Austral Islands, which are also rocky (with populations associated with brown algae). The typology of the French seabed is carried out under CORINE biotope[6] and is available for all the seabeds of metropolitan France and the French overseas departments and territories in the document "Vigie Nature" produced by the MNHN on behalf of the Ministry of the Environment.

2.2.2. *Sea-level fluctuations and coastline erosion*

Although the amount of water available on the planet (across all three phases) has remained unchanged, sea levels have varied greatly over the geological ages. Alternations between high and low sea levels due to glaciations have occurred over a period of time of the order of 20,000 years (Figure 2.3).

Since the beginning of the 20th century, with the production of greenhouse gases, the anthropic part of the sea level rise has increased. Given this phenomenon, the consequence will probably be a rise in water levels of about 50 cm before the middle of the 21st century. Unless human behavior changes, by the end of the 21st century, sea levels should be about 1 meter above current levels.

6 Database proposing a typology of natural and semi-natural habitats present on European soil, currently replaced by the European Union Nature Information System database for which it serves as a working basis.

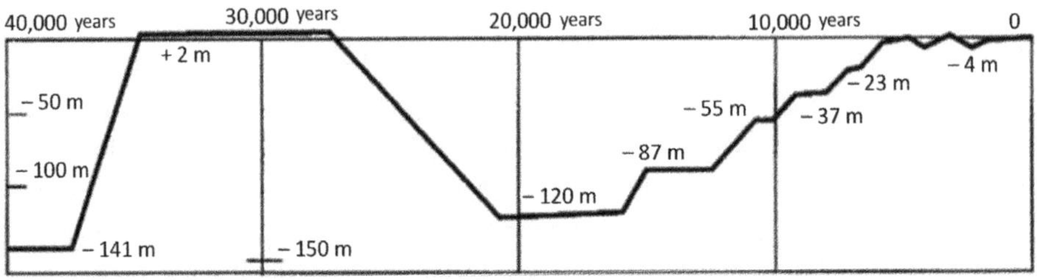

Figure 2.3. *Sea-level evolution over the past 40,000 years (source: Pinot (1998))*

It is clear that the evolution of the mean sea level impacts the coastline and the remodeling of the coastline. The sea tends to erode and transform the coastline until it is in balance with the new environmental conditions. The main factors of marine erosion are:

– abrasion of rock surfaces and alteration of rocks by mechanical, chemical and biochemical phenomena under the effect of tidal water and sea spray;

– modification of the geotechnical characteristics of loose sediments and modification of the slope balance;

– mixing, sorting and transport of loose materials released by the actions of living coastal creatures, of animal or plant origin.

Overall, erosion tends to cause coastlines to evolve according to their nature and their degree of maturity or regularization, resulting in: (i) coastlines barely reworked by the sea; (ii) coastlines evolving towards regularization; and (iii) regularized and stabilized coastlines.

This classification makes it possible to measure the degree of development of a coastline according to its degree of maturity. Coastlines that are not very developed or stabilized are under little threat and can be subject to sustainable development that does not threaten their balance. On the contrary, shorelines in the process of evolving are exposed to potential major transformations over the short or long term. It is prudent to let them evolve without covering them with human developments that will not be resilient and will be very difficult to defend.

The developer must therefore have this macroscopic vision of their development to validate the implementation of their work. Knowledge of the different types of coastline is therefore essential from a technical and environmental point of view.

2.2.3. *Rocky shores*

Rocky shores are governed by erosion. Either the rock in place is simply exposed, without having undergone strong erosive attack (these are the denuded coastlines), or the rock in place is attacked by erosion and retreats (these are the cliff coastlines).

The denuded coastal landscape is generally presented as a fairly low coastline where cultivated and built-up lands are located at low altitude and separated from the sea by modest projections. The terrestrial world, always emerged, comes to rest gently at the edge of the sea.

The interface between the terrestrial and marine worlds is materialized by a low wall, a hedge or a slope. Vegetation then abruptly gives way to bare rock, dotted with lichens and terrestrial plants, embedded in the fractures (Figure 2.4).

Figure 2.4. *The pink sandstone coast in Erquy, Côtes-d'Armor. Coastal landscape with denudation (photo: S. Pioch)*

In the presence of particularly resistant rocks that take a long time to erode and change little over time, the denuded landscape can adopt steeper slopes that retain a denuded coastline function but are also called false cliffs.

On the contrary, a cliff is a coastal slope fashioned from rocks of varying hardness, which evolves mainly in reaction to the attack to its base by the sea. The latter has two fundamental aspects: (i) the undermining of the rock in place, which causes the upper part of the slope to become unbalanced and fall onto the foreshore; (ii) the clearing of debris by the sea, which dissolves it or transports it elsewhere. The cliff cannot know equilibrium and exists only because it is retreating. A cliff only stops retreating if the sea stops clearing its base (Figure 2.5).

Rocky shores are solid substrates that can accommodate sustainable infrastructure and facilities only if the coastline is low enough and the rock is hard enough that they evolve into denuded shorelines. This provides a reliable and solid substrate, with easy access to the sea and with sufficient elevation to ensure that development is protected from the sea and easy access to the interior. These advantages are multiplied in the case of coastlines alternating between bare coasts and beaches.

Figure 2.5. *Cliffs at Étretat in France (photo: J.-C. Souche)*

2.2.4. *Soft coastlines*

Soft coastlines are of increasing economic interest with the development of seaside tourism and water sports. However, their natural equilibrium is fragile and unstable, and a small modification brought to soft coastlines by an infrastructure can be enough to generate significant unexpected and difficult changes. In all cases, it is necessary to avoid designing developments on an *ad hoc* basis according to the observed degradation, without any overall reflection. The greatest problem posed by their development is preserving their interests while opening them to the public. The most relevant option consists of preserving the natural appearance of the sites by making only discrete developments and minimizing the damage caused by frequentation.

These shorelines are governed and regulated by the solid transport of their constituent materials. Whatever the size of the non-cohesive sediments (sand grains, gravels or pebbles), they are governed by the mechanical forces acting on them individually. Only the level of energy required to mobilize them differentiates them according to their size and weight. It should be noted that some loose shorelines formed of fine sediments (silts or clays) have a reduced mobility as the cohesion forces give them an overall mechanical behavior.

Soft shorelines have a subtidal zone (foreshore), always submerged, which exchanges sand with the beach. Its depth varies from a few meters in sheltered areas to a few dozen meters in particularly exposed areas. The slope of the foreshore faces out to sea and its isobaths are roughly parallel to the crests of the prevailing swells. The boundary between the foreshore and the ordinary seabed is generally marked and evolving. The ordinary seabed is little affected by current swells, but is stirred up by exceptional swells. Depending on the quantitative balance of the inflow, the shoreline expands or retreats during seasonal exchanges. The profile of the foreshore generally presents slopes of the order of a few percent for sandy seabeds.

The emerged part is the beach or intertidal zone (Figure 2.6), part of which is covered and uncovered with the tide (the foreshore). Beach sediments are characterized by their mobility under the action of swells and range from sand (finer particles) to pebbles (larger particles). The profile of a beach is the assembly of a series of slopes, each of which corresponds to a dimension of the sediments.

At the back of the beach, we may find a dune which is an accumulation of sand calibrated by its mode of transport, the wind, which forms an elevation of the relief out of reach of the usual waves. There can only be a perennial dune if the foreshore functions with a net contribution of sediments from the open sea to the land and if the wind cannot immediately carry away the sand that it has just deposited. The presence of obstacles, natural or artificial, is thus an essential element of the permanence of the dune. A dune is stable, but not immobile (Figure 2.7).

Figure 2.6. *Profile of a beach exposed to prevailing westerly swells at Étel in Morbihan (photo: J. Bougis)*

Figure 2.7. *Beach and dune of Espiguette in the Gard (photo: J. Bougis)*

2.2.5. *Fine material shorelines*

The term "silt" refers to a sedimentary mixture whose cohesive behavior is imposed by the presence of fine mineral elements, called "fines", sometimes combined with organic elements. A silt is usually composed of sandy or even gravelly elements, sometimes even pebble-sized, but it is the presence of lime and clay that gives it its cohesion. This last intrinsic property gives it mechanical characteristics.

A coastal mudflat can only form in areas that are open enough to the sea for sediment-laden water to reach, and sheltered enough from swells and currents for fine material to settle and consolidate (Figure 2.8).

Figure 2.8. *Coastal mudflat in the cove of Sauzon (photo: J. Bougis)*

Mudflats generally form in areas sheltered by obstacles (bays with narrow entrances, estuaries, depressions blocked by sandy spits, etc.) or within shallow areas spread over large surfaces where the swell breaks due to the depth and loses its energy at the entrance to the area.

2.2.6. *Coral bioconstructed coastlines and mangroves*[7]

A coral reef is a mineral construction made of calcium carbonate that is bioconstructed by living beings, mainly corals of which there are many species. Coral reefs differ from the accumulation coasts because they are of biological origin.

Reefs take different forms depending on the stage of their evolution: (i) a reef bed forms a coral colony that develops into a reef platform, a flat area located close to the coast; (ii) when the flat area moves away from the coastline, it forms a fringing reef that creates a depression between it and the coastline (e.g. Reunion Island); (iii) in more developed stages, it can form a coral reef that breaks off at the right of the passes with the ocean, sheltering a lagoon ranging from a few to tens of meters in depth (such as the Ile de Mayotte); (iv) it can finally form an atoll that is a coral crown delimiting a central lagoon all around the island.

A mangrove is an ecosystem comprising a set of plants such as the mangroves that grow on the foreshore of tropical regions in calm and shallow areas. Mangroves stabilize coasts against erosion and thus reduce the risks of natural disasters, such as cyclones or tsunamis or the effects of global warming. Mangroves are very productive ecosystems in terms of biomass.

Our oceans contain no less than 600,000 km^2 of coral reefs, about 600,000 km^2 of seagrass beds and 152,000 km^2 of mangroves (Figure 2.9). These ecosystems are rich in diversity: the total animal and plant species richness of coral reefs has been estimated at 93,000 species (Reaka-Kudla 1997). In some regions (e.g. the coral triangle in Indonesia), 500–600 different species have been described. It is estimated that one-third of marine species (fauna and flora) live and depend on these coral reefs (Figure 2.9).

These ecosystems therefore present a considerable socio-economic challenge for coastal populations, particularly with regard to coastal protection by reducing damage to coastal infrastructures, following repeated wave action or high intensity climatic events. They have a strong attraction as regards tourism and provide a source of income related to the production of biomass associated with them, thus serving to support the coastal economy. For example, fishing activity, whether commercial or

7 This section was co-written with Aurore Leocadi, oceanographer.

recreational, is very strongly linked to their presence. The same is true for seaweed, which is used everywhere in industry (toothpaste, cosmetics, animal and human food, etc.).

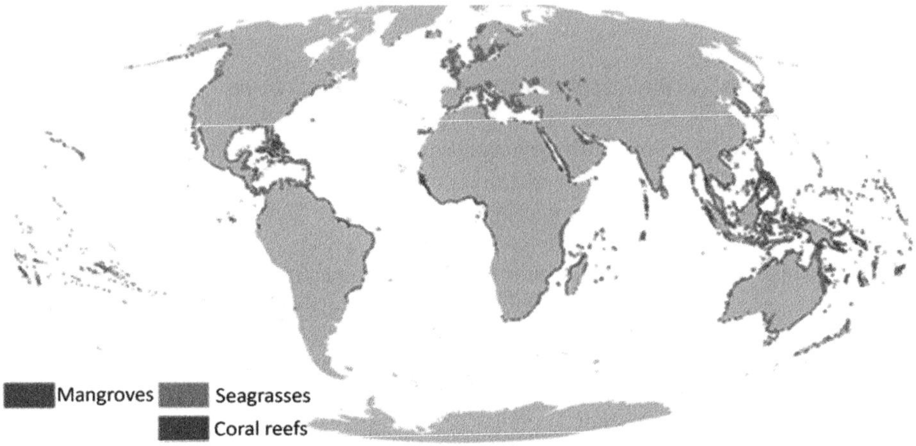

Figure 2.9. *Seabeds colonized by coral reefs (red), seagrass beds (green) and mangroves (blue) in 2018 (source: WCMC). For a color version of this figure, see www.iste.co.uk/pioch/marine.zip*

These ecosystems have a fragile natural balance, and they are currently under serious threat: 50% of mangroves have been destroyed, 60% of coral reefs are threatened and a decrease in the surface area of seagrass beds has been noted. In 2016, according to Hughes *et al.* (2017), only 8.9% of the 1,156 reefs comprising the Great Barrier Reef escaped a bleaching event[8] (Figure 2.10).

Humans are directly or indirectly linked to the health of these ecosystems. Whether they are anthropic or natural in origin, numerous pressures can irreversibly modify their ecological state to a greater or lesser extent. Generally, they accumulate and multiply their effects within the same ecosystem through a synergistic effect.

The resilience of these ecosystems is highly subject to the impacts of human activities (wastewater discharges, agricultural, industrial and/or port activities, activities related to the use of the sea, etc.).

8 Bleaching is a phenomenon of decay that results in discoloration. It is due to the rise in water temperature, which leads to the expulsion of symbiotic algae that give the coral its color and nutrients (see https://www.futura-sciences.com/planete/actualites//rechauffement-climatique-grande-barriere-corail-nouveau-victime-blanchissement-62231/).

However, development projects remain one of the main threats to the health of these ecosystems. The consequences can be multiple: habitat destruction, landscape degradation, changes in currentology, hypersedimentation and acidification (Reaka-Kudla 1997; Hughes *et al.* 2017).

Figure 2.10. *Coral bleaching in 2016, Reunion Island (photo: MAREX/J.-B. Nicet)*

2.2.7. *Assessment*

Initial analysis therefore consists of classifying coastlines and highlighting the vulnerability of some of them to erosion and sea attack. It also makes it possible to identify valuable natural sites where it will be decided not to develop. This is the first step of the ARO triptych that we recall here, "Avoid", for environmental and/or technical reasons.

This double identification for the correct implementation of the development within the coastline is essential to determine the suitability of maritime developments with their future geomorphological context. The successful integration of this infrastructure with the coast also depends on the functions of these structures, which are classified into two families: coastal civil engineering for coastal defense and port structures.

2.3. Coastal defense works

Coastal defense consists of the establishment of infrastructures capable of responding to attacks from the sea. There are several systems of structures that can be used for coastal defense. If longitudinal actions are dominant, the coast will be protected by means of cross-shore structures that are designed to interrupt longitudinal transport. If perpendicular actions to the shore are dominant, coastal defense will be ensured by longitudinal structures of the top of the beach or even positioned in the sea (breakwaters).

Cross-shore structures are essentially groins that stop the littoral transit: (i) short groins act on cobble beach transport; (ii) long groins extend down to the breakwater bar and also act on littoral flow transit. They are used on sandy beaches.

Longitudinal structures are made up of breakwater batteries which, by limiting the energy of the swell in the surf zone, modify the conditions of erosion and allow the creation of tombolos. At the top of the beach, the "wave breaker" walls fix the coastline and protect the backshore.

Finally, we should not forget the methods of shoreline and dune maintenance, beach nourishment and restoration of coastal transit.

A shoreline defense scheme is often composed of a combination of different types of structures and is accompanied by beach top maintenance methods. It should also be noted that the protection of a shoreline site must begin with its non-degradation.

Under these conditions, the cessation of material extraction (quarries, pebbles, sand, coral, etc.), apart from those used for coastal management, and the prohibition of certain human activities that are particularly erosive (mechanical sports, mass passages, mass tourism, etc.), are an essential prerequisite.

2.3.1. *Main transverse structures: groins and jetties*

Groins are mainly classified according to their permeability, length and height. The materials used to build them are mainly rockfill, steel, concrete and, more rarely, wood. The choice of a type of groin depends mainly on the tidal range and the wave energy (height) on the site.

– *Rockfill groins*: these are made from shell rockfill blocks, with or without a core depending on the desired permeability. The blocks must be sampled for stability against hydrodynamic actions. In cases where watertightness is required, the geotechnical rule of filters between the core and the armor should be respected (Figure 2.11).

– *Steel groins*: these are constructed of metal sheet piling. They can be made permeable by means of cut-outs in the steel. Sheet piles can be single or double. In cases where the stability of a sheet pile wall is insufficient, it is preferable to build closed cells and then fill them with sand or pebbles.

Figure 2.11. *Rockfill groins, Espiguette beach, Gard, Mediterranean Sea (photo: J.-C. Souche)*

– *Geotube groins*: a new groin construction technique that consists of filling a geotextile coil with sand from the site or with an input material. When adapted to the site conditions, these solutions are economically advantageous.

– *Wooden groins*: these are made up of wooden or metal frames, planks, piles or posts. The boards that are slid into grooves within the posts can be separated by a gap (permeable groin) or joined together (watertight groin). To ensure a reasonable life span, the wood must be rot-proof or treated. These solutions used in the past are becoming rare nowadays.

– *Jetties*: structures built in the water in order to direct and confine a river or a marine current in a channel and avoid its silting up by the littoral transit. Located at the entrance of a bay or a river, the role of jetties is to protect the channel, on the one hand, against the waves and the agitation which results from them, and, on the other hand, against the crossing currents. When they are located at the mouth of a river, or at the entrance of a port, at the level of a barrier beach, jetties also contribute to fixing the opening and controlling the littoral transit on adjacent beaches. They then fulfill the role of groins.

2.3.2. *Main longitudinal structures: breakwaters and upper beach structures*

The function of structures parallel to the shoreline positioned on the front beach, generally called breakwaters, is to directly oppose the swell and the circulation of currents perpendicular to the shoreline. They therefore oppose solid movements in the beach profile.

The activity of breakwaters remains limited to their vicinity because they act in two ways: (i) they reduce the energy of the waves, by reflection and breaking of the swell; and (ii) they spread the energy of the swell, by diffraction around the ends and refraction on the bottom.

Breakwaters can be emergent or completely submerged:

– emergent structures create a zone of relative calm favorable to sedimentation between them and the coast;

– submerged structures limit the volume of water that crosses them and constrain the swells that exceed a camber limit.

Today, breakwaters are essentially constructed as rockfill embankments. They are provided with a core of quarried rock and may be fitted with geotextiles. Rock armor shells can be arranged to varying degrees (Figure 2.12).

Figure 2.12. *Rock breakwater, port of Bastia (photo: J.-C. Souche)*

In some cases, shells made up of carefully arranged prefabricated concrete blocks are used. The most successful process is the CLI system, based on ACCROPODE or ECOPODE blocks, developed by Artelia. The laying plan allows these blocks to work together in a global way and avoids the justification of each block with respect to the swell. The result is single-layer shells that are thinner, which is economically interesting and minimizes the footprint of the dikes on the sea floor (Figure 2.13).

Other breakwaters are constructed using geotubes covered with a rockfill or non-rockfill shells (Figure 2.14).

Figure 2.13. *Ospedaletti-type ECOPODE™ and ACCROPODE™ II blocks (photo: E. Cunge)*

Figure 2.14. *Geotube® breakwater (source: commercial documentation – TenCate Geosynthetics®)*

Beach nourishment structures are designed and dimensioned according to the forces generated by the sea and the supporting forces generated by the backshore they protect. Resistance to the forces due to the sea comes from their weight, shape and the mechanical resistance of their structure and coverage.

The structures can be simple walls separating the beach and the city without changing the relief, as is the case at Narbonne-Plage (Figure 2.15).

Figure 2.15. *Narbonne-Plage, low wall at the top of the beach to contain the sand on the beach side (photo: J. Bougis)*

However, most often, the levels are different between the beach and the backshore. The structure then becomes a riprap. The oldest are the masonry ripraps (Figure 2.16). These have a very long life span and are barely threatened by overly low scaling of the facing, poor implementation or poor maintenance (repointing, clogging).

Figure 2.16. *La Flotte-en-Ré, Louis-Philippe-era masonry riprap (built in the middle of the 19th century) (photo: J. Bougis)*

Concrete structures have been built for many years (Figure 2.17). They have a very long service life, unless defects are found in the execution of the work (coating defects, poor quality of concrete or poor placing).

Figure 2.17. *La Ciotat, wave protection wall of the shipyards at the top of the dike (photo: J.-C. Souche)*

Structures made up of natural rockfill or artificial blocks are made in the same way as the embankment dams. They are currently widely used.

2.4. Port structures

2.4.1. *Criteria for establishing a port*

A seaport is a facility located on a sea or ocean coast that is intended to accommodate ships during loading and unloading operations and for refueling and repairs. A port is also a place of stay, a particular example being marinas.

In general, a port is targeted for a particular use, adapted to given vessels, and thus we distinguish between commercial ports, whose infrastructures are adapted to the various goods transported (containers, bulk, oil, etc.), fishing ports, marinas and military ports.

Knowledge of the coastal environment and the coastline is essential to correctly understand the setup of a port when it is being built. Indeed, a host site is characterized by: (i) geographical conditions; (ii) economic criteria for construction and operation; (iii) the fleet of vessels to be accommodated; and (iv) operating principles. These hypotheses are established for a service life of several decades without prejudging their future evolutions, which will require the requalification or redevelopment of the infrastructures.

2.4.2. *Geographical location*

A port is said to be "natural" if it is located within a roadstead, a cove, a bay, an estuary or a fjord that offers a natural protection against the elements. Most of the ancient ports created on our coasts were within natural ports.

Conversely, it they also be external and artificial. In this case, the port will be built *ex nihilo*, either by creating a dock by dredging and excavating the coastline or by building dikes and jetties in front of it. This is called a coastal port.

For example, the large seaport of Marseille Fos-sur-Mer includes, among other things, a historic natural harbor –"the old port" – and a coastal port to the west (Figure 2.18).

Figure 2.18. *View of the port of Marseille, old port and western basins (photo: B. Sigros)*

Inland ports (e.g. the ports of Rouen or Nantes) are established on rivers and are connected to the sea by a maritime canal. They sometimes include an outer harbor.

2.4.3. *Economic criteria*

For decades, the maritime transport sector has been undergoing major transformations. International trade is developing at a very fast pace, in an extremely competitive context where maritime transport prices are being driven down. As a result, shipowners' margins are being squeezed and a race to cut costs is underway. Giant groups and alliances are being formed and each operator is pursuing a strategy of concentration and acceleration of transport. For example: the average size of container ships is increasing, minimal time savings are being exploited, the most direct routes are being prioritized and ports whose hinterland is highly populated and economically very active are preferred.

Depending on the performance of ports, much added value can be created or lost within them. Indeed, shippers are concerned about the entire journey of their goods, from the place of loading to the final consignee. Many of their concerns are in line with those of the shipowners; they want to minimize their inventories and offer impeccable quality and are looking for both a good price and fast transport, combined with high reliability. Only a few ports will be able to meet these requirements and a dense and developed hinterland, an excellent quality of service or a favorable geographical position on main world trade routes have become indispensable.

A port assumes economic needs justifying its creation. These needs are to be defined at time t_0 but also in terms of future development. Indeed, a port has two essential functions:

– first, a shelter for boats, with all the functions previously described;

– it is also a place for the exchange of goods or people between the sea and the land. As previously mentioned, without its commercial hinterland, a port is nothing.

Therefore, in addition to the physical geography criteria discussed above, consideration of economic geography criteria is essential in the decision to establish a port. The following flow chart (Figure 2.19) summarizes the criteria for establishing a port.

Based on the missions of the ships that use the port, it often takes the name of the main function although it usually fulfills several functions.

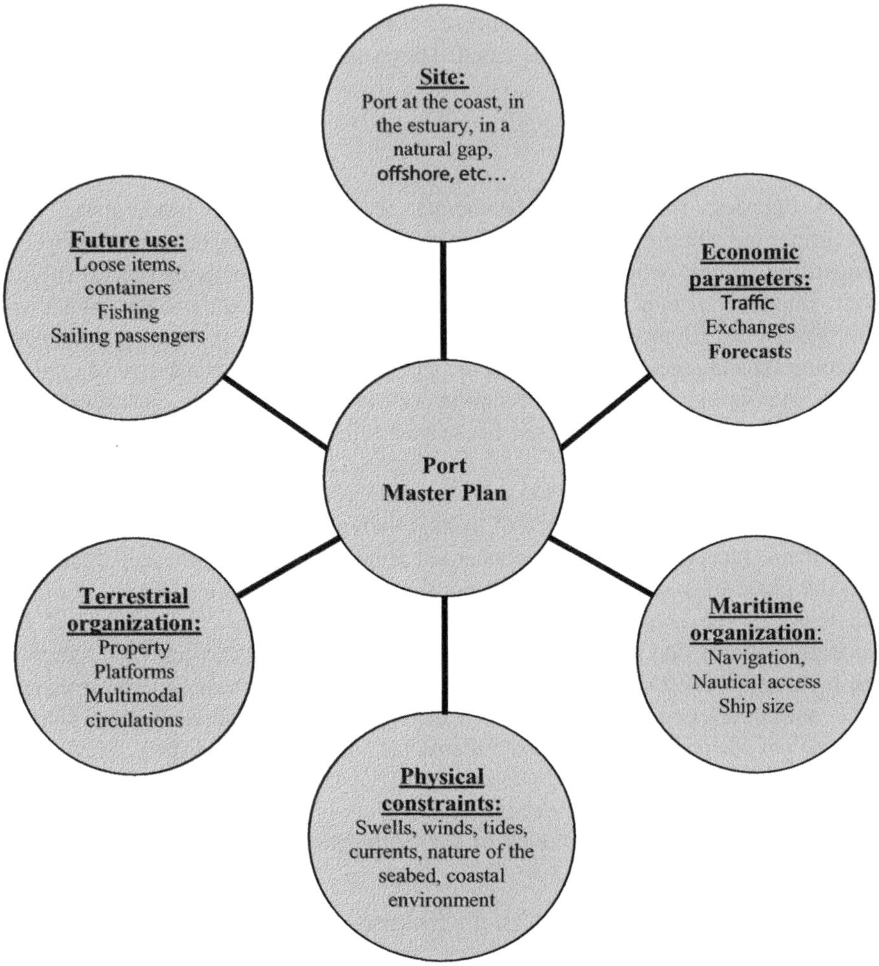

Figure 2.19. *Criteria used to establish a port*

Thus, a distinction can be made between military ports (anchorage, repair, bunkering, maintenance, etc.) and commercial ports (passenger, cargo, bulk cargo, containers, Ro-Ro, fishing ports and marinas) (Figure 2.20).

This distinction will be important because operators (the army, autonomous ports or local authorities of various sizes) do not have the same culture, nor the same constraints with regard to the new concepts of eco-design, which we will develop later on.

a)

b)

Figure 2.20. *a) Example of a marina, Grande-Motte harbor; b) Example of a commercial harbor in Bastia (Ro-Ro docks for ferries and passengers) (photos: J.-C. Souche)*

2.4.4. *Operating principles of ports*

At the start of the creation of any port infrastructure, there is a site which, in the most general case, is virgin; the geographical and climatic conditions specific to this site are essentially characterized by a morphology of the coast and tidal (Figure 2.21), wind, current and swell regimes.

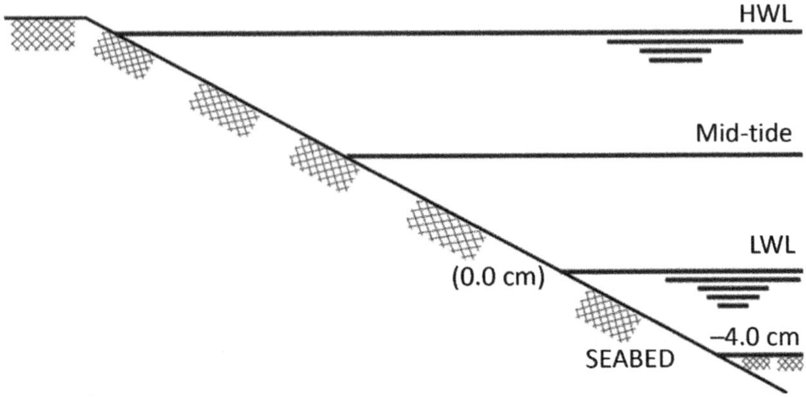

Figure 2.21. *Example of a typical cross-section of a coastline to be developed*[9] *(source: B. Sigros and J.-C. Souche, marine works course materials, École des Mines d'Alès, 2020). For a color version of this figure, see www.iste.co.uk/pioch/marine.zip*

The draft and the dimensions of the ships received will have an impact on the depth of the port and consequently on its operating principles. The marine environment of basins and their structures will be particularly important if a connection with the marine environment is sought for environmental purposes.

Three main types of harbor operation are possible, depending on the characteristics of access to the harbor and the desired moorings: (i) the dry dock; (ii) the deepwater port; and (iii) the wet dock and its variant, the constant level wet dock.

2.4.4.1. *Dry dock*

In this type of structure, the only objective is the protection of boats from currents and swell. For this purpose, only a protective structure is built (Figure 2.22). Access to the harbor at low tide is not possible.

Protective dikes can therefore be established at a short distance from the coast; they are not very high and the result is a moderate construction cost.

9 LWL: lowest water level, which is also the zero point of nautical charts; HWL: highest water level.

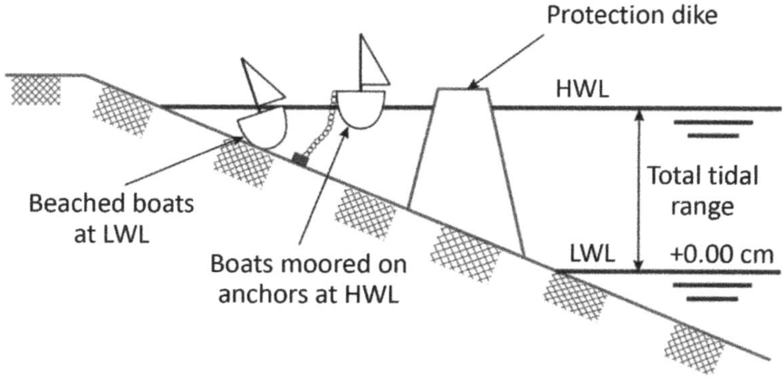

Figure 2.22. *Example of the operation of a dry dock (source: B. Sigros and J.-C. Souche, marine engineering course materials). For a color version of this figure, see www.iste.co.uk/pioch/marine.zip*

Figure 2.23. *Bassin du Roy, boats at anchor, port of Le Havre (photo: J.-C. Souche)*

The "comfort" of the boats is rudimentary: at each low tide, boats run aground; otherwise, access to the coast is by means of a small tender boat. In hydraulic terms, the port and its structures are subject to the tidal regime in the same way as the open sea (Figure 2.23).

2.4.4.2. *Deepwater port*

This type of harbor is also called a "tidal basin". In this type of port, the basin is dug at a level that allows boats to stay afloat whatever the sea level; if a draft (including safety margin) of 3.50 m is required, the basin must be dredged or excavated up to the coast at a depth of 3.50 cm.

Protective structures should be established on natural beds at the appropriate shoreline to allow access for boats in all weather conditions, regardless of sea level (Figure 2.24).

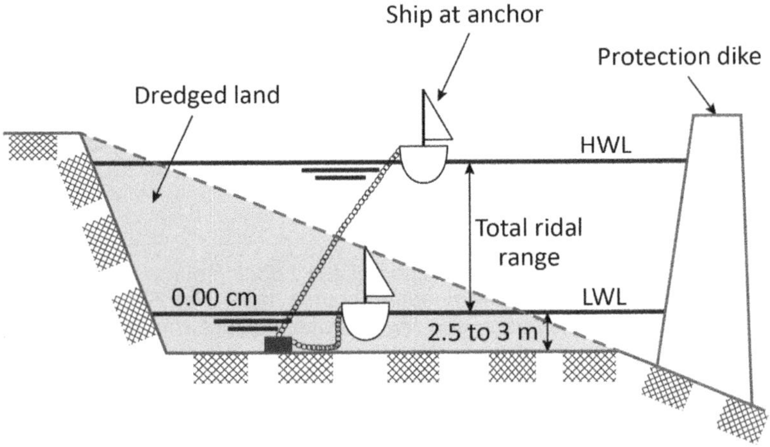

Figure 2.24. *Example of the operation of a deepwater port (source: B. Sigros, J.-C. Souche, maritime course materials). For a color version of this figure, see www.iste.co.uk/pioch/marine.zip*

In contrast to the dry dock, the "comfort" and protection of the boats is at the highest possible level (Figure 2.25). On the contrary, the investment cost of the harbor is maximum because its construction requires a lot of earthworks and dredging, as well as very high protective structures, which are therefore subject to higher swell loads due to the great depth of water.

Figure 2.25. *Example of the marina of La Rochelle (photo: J.-C. Souche)*

2.4.4.3. *Wet dock*

The port is a closed basin inside which the water is held at a minimum level set according to the level of a weir that allows boats to enter and exit. The level of the sill determines both the depth to which the basin must be dredged and the length of time that boats may or may not enter the harbor during each tide. A high sill level does not require much earthwork, but will limit the range of access to the harbor, whereas a low sill level will require more earthwork, but will increase accessibility to the harbor; in this case, it is similar to the wet dock case described earlier (Figure 2.26).

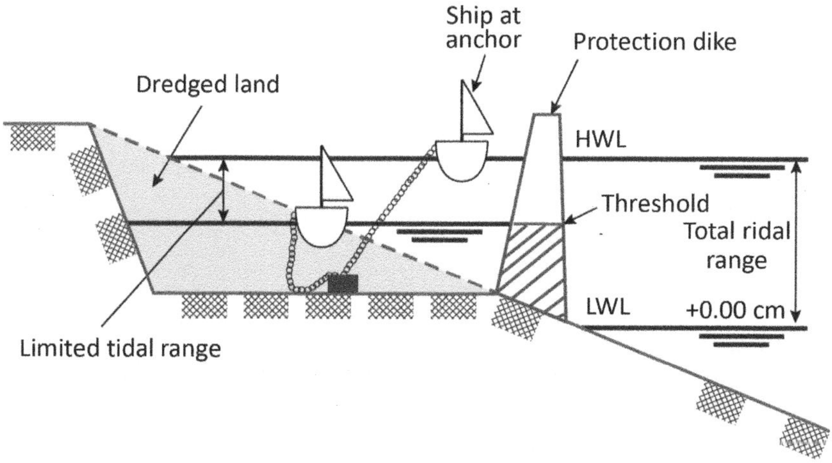

Figure 2.26. *Example of the operation of a wet dock (source: B. Sigros and J.-C. Souche), maritime course materials. For a color version of this figure, see www.iste.co.uk/pioch/marine.zip*

Accessibility to the wet dock can be improved without increasing the volume of land to be dredged by moving the entrance sill seaward (Figure 2.27).

The constant-level wet dock is a variant of the above case, which allows the significance of dredging to be minimized by maintaining the water level in the wet dock at a value generally close to mean high tide. Access to the port is necessarily controlled by a lock (a "simple lock" or "lock chamber") (Figure 2.28). Depending on the water level selected, it may be necessary to compensate for water losses caused by the entry and exit of vessels by means of a pumping device.

Figure 2.27. *Example of the Sables-d'Olonne wet dock (photo: J.-C. Souche)*

Figure 2.28. *Example of the Saint-Malo wet dock (photo: J.-C. Souche)*

2.4.4.4. *Assessment*

An understanding of the hydraulic functioning of the port is necessary for the technical dimensioning of the structures. Moreover, connection of a structure with the marine ecosystem depends on its immersion time and thus on its exposure to tides, which depends on the operating principle of the port.

2.4.5. *Typology of port structures*

Port structures have quite diverse functionalities because they meet different design criteria and numerous execution constraints that will influence the structural choices.

Coastal or harbor defense works and open sea works are usually called external works. The harbor structures related to the service (mooring, berthing, loading, etc.) are generally called internal works.

A port is composed of a maritime area comprised of several docks or basins, themselves comprised of quays connected to a terrestrial section that enables multimodal links with other means of transport (road, rail, etc.). In this chapter, we limit ourselves to maritime structures directly connected to the marine environment.

The port also includes structures intended for the maintenance of ships (dry slipways, dry construction areas) or dedicated to the navigation of ships (locks). The port, its docks and its structures must have a depth adapted to the ships to be accommodated, which may require earthworks or dredging in order to create or maintain a sufficient depth.

Finally, some maritime structures will not be located within the port area, nor will they be located on the coast to defend it. We are referring here to offshore and open sea works, such as an outfall, for example.

Table 2.2 summarizes the main types of marine structures, their dominant functions, common construction materials and brief structural or construction principles.

For marine structures, knowledge of the data, constraints and assumptions is essential, enabling us to propose a design adapted to the geographical, oceanological and technical environment. This simple aspect of things is already complex and requires strong expertise to propose adapted solutions.

If, from now on, the eco-design dimension is to be taken into account, knowledge of coastlines and the natural environment as a habitat for living organisms becomes unavoidable. Indeed, the design of the work becomes completely linked to the problematic of the neighboring coastline and its capacity to welcome or exchange biodiversity.

The marine engineer will therefore be obliged to have a global vision, which will be necessary for good hydraulic design as well as for the good definition of the relationship between the structure and the environment necessary for the eco-design of the structures. The design of maritime structures will become a direct consequence.

Types of structures	Features and summary description	Usual materials
External works (protection of the harbor from the effects of the swell)		
Breakwater	Longitudinal defense structure to protect the port from the actions of the sea by damping the swell	Rockfill Concrete blocks Geotubes
Groins	Defense structure perpendicular to the coast, to protect it from the actions of erosion and solid transport	Rockfill Geotubes Steel sheet piling
Dike, jetty, breakwaters	Protection of port areas against swell and currents	Embankments and rockfill Embankments and concrete blocks (cubic blocks, tetrapods, accropodes, etc.) Reinforced concrete caissons on ballast Concrete block docks
Heavy pontoon breaker	Floating structure making it possible to protect the port infrastructures from the swell	Pontoons from reinforced concrete or metal held on steel piles or concrete anchors
Inland port structures (with port service functionalities)		
Wharfs	Docking structure enabling the loading and unloading of ships	Concrete deck or mixed structures on massive concrete caissons or gabion piles of sheet piling
"Danish" wharves (wharves on piles) landing stages, wharf	Structure on piles constituting a wharf (if perpendicular to the coast)	Concrete or steel deck on steel piles or reinforced concrete
Screen docks or caisson docks	Continuous structure on a platform allowing the docking of boats and their loading or unloading	Reinforced concrete caissons Steel sheet pile gabions Sheet pile walls or combi walls (sheet piles and piles) Reinforced concrete screen of the cast wall type
Pontoon	Floating structure for mooring and berthing boats in a port	Metal or reinforced concrete, held on metal piles or concrete anchors Aluminum for boating or fishing
Locks/gates	Opening allowing a boat to cross hydraulic drops and/or keep a basin afloat	Reinforced concrete infrastructure, equipment generally in steel

Types of structures	Features and brief description	Usual materials
Inland port structures dedicated to ship maintenance		
Dry construction docks, dry slipways	Watertight basin which can be dried or immersed enabling the maintenance and repair of boats Lock gate managing the access and watertightness of the dry dock	Non-reinforced concrete, reinforced concrete Reinforced concrete, prestressed concrete, steel
Slipway	Inclined plane enabling the dry docking of a boat for maintenance or for launching	Reinforced concrete
Self-lifting platforms/boat elevators	Mobile apron for the drying of vessels for maintenance and/or mobile equipment using a dock for the drying of vessels for maintenance	Reinforced concrete infrastructure, steel equipment
Open sea structures of various functionalities other than protection		
Anchors, weights	Underwater structures for the anchoring of a ship or a floating infrastructure	Reinforced concrete
Artificial reefs	Underwater structures for benthic or pelagic fauna	Reinforced concrete, steel, various materials
Pipes and outfalls at sea	Submarine pipelines with the purpose of transporting a fluid or channeling an effluent flow off the coast	Pipes generally in polyethylene, weights and anchors in steel and anchors in steel or in concrete
Offshore structure (for the record)	Structure located at sea in an oil exploitation area or for a wind farm	Reinforced concrete or steel infrastructure

Table 2.2. *The main structures used in maritime sites*

2.5. Design approach

2.5.1. *The design methodology*

A project always starts with the definition of the need by the project owner through the development program, usually in response to socio-economic considerations.

In response, the designer proposes an architectural and technical response to the project, the feasibility of which must be verified, both technically and financially, as well as with regard to the various regulations or the available land, etc.

The very first step for the designer is to consider the development in a macroscopic vision within the coastline, in order to choose the appropriate implementation or, if it is fixed, in order to understand the issues to which the development will be subjected (solid transport or hydrodynamic action, for example). In addition, the designer must also understand the operating principle of the port. These aspects will have an impact on the design of the structures to be built. Also at this initial stage, the designer must draw up a list of the data and hypotheses of the project and establish the constraints to be taken into account.

Next, the designer must choose a type of work and a design for the work. This is generally based on research of the local techno-economic optimum specific to the development to be built. The design is supported by calculations that allow the design to be refined and lead to specifications for the materials to be used and the methods of implementation (Figure 2.29). The design phase therefore ends with a "project" that specifies the description, the cost of the work and the estimated completion time. It also specifies the quality standards to be achieved. The project serves as a basis for the work contracts of the companies for the realization of the maritime infrastructure.

The next logical step is phase 3, which consists of the construction of the structure via execution of the works. It ends with the acceptance of the works and the transfer of ownership to the project owner (and/or its operator), as well as the start of the operation of the infrastructure, its monitoring and maintenance.

This final phase, the operation, is the longest in time, lasting for decades. It is during this phase that the structure will be subjected to the elements and the actions of service. It is at the end of this phase that durability can be verified. It goes without saying that the operation must be completely consistent with the calculation and load assumptions of phase 1. This is the difficulty of designing a maritime structure: foreseeing what will happen to the infrastructure in service (for 30, 50 or 100 years, depending on the case), while many uses will evolve during this period. This is the technical and economic dimension of things, but it is not the only one.

Indeed, the development must be integrated into its environment, by being respectful of humans, landscapes and natural environments. The design must be concerned with saving space, sparing species and limiting water, air or soil pollution.

To verify these requirements, an environmental consulting firm must carry out an impact study, which is an essential step in the environmental assessment of development projects. This approach is intended to help the promoter integrate environmental concerns into the design phase of a development (but too late, as recalled in Figure 1.2), in order to inform the services called upon to decide on the appropriateness of authorizing its realization and to inform the public by involving them in the decision-making process (Article L. 122-1 of the Environmental Code; European directive of June 27, 1985).

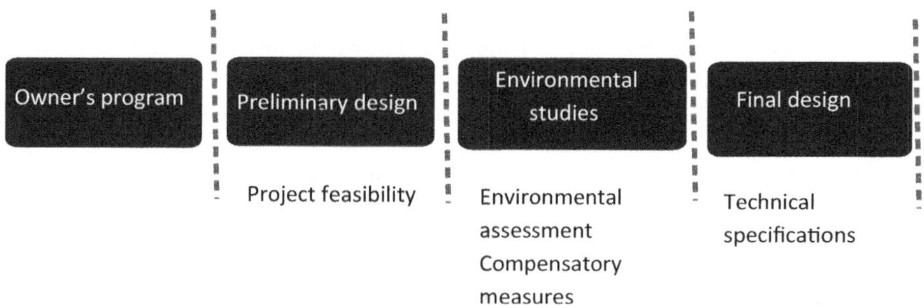

Figure 2.29. *Design phase execution process of a development*

Technical and environmental studies complement each other and are often carried out by different actors, which does not always allow for the compilation of an integrated vision of the development.

Therefore, achieving these objectives requires a good understanding and appreciation of the data, constraints and assumptions of the facility. They must be an accurate reflection of what the service conditions will be during the entire life of the structure, some of which are listed below:

– topographic and bathymetric data near the structure and offshore. These data are generally not difficult to acquire[10], but they require resources and must be in the possession of the designer from the beginning of the study. They must be recent (bathymetry in particular) because the solid transport of the coastline makes them evolve at different rates;

– ocean data, including tides, currents, wind, swell;

10 Their acquisition can be costly however!

– service data: project vessels, handling tools and operating conditions (berthing speeds, mooring plans). These operating constraints must be known in order to completely define certain actions, such as the berthing and mooring of ships at the quay, or the loads of cranes and loading/unloading gantries;

– building materials data and assumptions;

– geotechnical data that will lead to the determination of the foundation conditions: (i) knowledge of the site stratigraphy and geological conditions; (ii) determination of the mechanical and geotechnical characteristics of soils by means of a reconnaissance campaign[11]; (iii) evaluation of the constraints in the provisional phase (stability of soils, neighboring slopes, settlements, etc.);

– the existing structures, the exact characteristics of which (foundations, structure, dimensioning hypotheses) must be known in order to adapt the usable construction methods and the connections of the new structures to be built to the old structures.

Taking into account all these elements often leads, before any justifying calculation, to one or more of the technical solutions being either imposed or eliminated.

In some cases, a geological context or hydrodynamic conditions may be so impactful that their significance may influence the selection of a construction site. In the future, we could imagine that a similar conclusion might be reached in light of ecological data from a natural landmark that would steer the siting of a development elsewhere.

In the rest of this book, the authors will detail the data that must be correctly understood from the beginning of the design studies in order to mitigate the risk of disaster later.

2.5.2. *Ocean data*

2.5.2.1. *Tides*

Oceans and seas are liquid masses always in movement under the effect of different cosmic (Moon and Sun) or terrestrial (wind and atmospheric pressures)

11 The acquisition of geotechnical data is always expensive because it mobilizes important resources in a maritime environment, where everything is more complicated than on land, an example being access to the permissions for a future structure to be built on the sea. Moreover, the choice of the tests to be carried out itself depends on the type of structures to be built and the construction options. The definition of an adapted and optimal geotechnical campaign thus requires a certain level of experience in the marine civil engineer.

activities. This results in wave motions, such as tides and swell and general oceanic currents of varying uniformity.

The tides are an oscillatory movement of the sea level generated by the attraction of the Moon and the Sun on the fluid particles of the oceans. These astronomical tides are the best understood and are the result of the diurnal motion of the Earth's rotation around the axis of its poles, which causes the fluid particles of the oceans to move in the generating force field of heavenly bodies, such as the Moon or the Sun. When the Sun, the Earth and the Moon are approximately aligned (new or full moons), the generating forces created by the Sun and the Moon are combined; the resulting force is at its maximum and this is the period of spring tides. At the periods of quadrature, the actions generated by the two bodies (Sun and Moon) are opposed and the resulting force is minimal; this is the period of neap tides (Table 2.3).

Coefficient	Type of tide	Tidal range in Brest
120	Highest astronomical tide (HAT)	7.32 m
95	Mean high water (MHW)	5.80 m
70	Mean sea level (MSL)	4.27 m
45	Mean low water neap (MLWN)	2.75 m
20	Lowest astronomical tide (LAT)	1.22 m

Table 2.3. *Main tides and their associated coefficients and tidal range in Brest (source: ROSA (2000))*

The lunar day, like the solar day, is approximately 24 hours long, so tides are semi-diurnal and alternate between high and low water approximately every 6 hours (Figure 2.30). The periods of high water appear every 14.5 days, or half a cycle of the lunar revolution around the Earth. Forecasts are published on the network of the French *service hydrographique et océanographique de la marine* (SHOM)[12].

In addition, there are non-astronomical causes of sea-level variations; they are meteorological in nature and generate over- and under-runs. They are due to: (i) high or low barometric pressure; (ii) winds (movements in bodies of water); (iii) currents; and (iv) storm swells (possibly cyclonic). They are often referred to as meteorological or barometric tides when only the effects of atmospheric pressure variations are involved.

12 https://maree.shom.fr/.

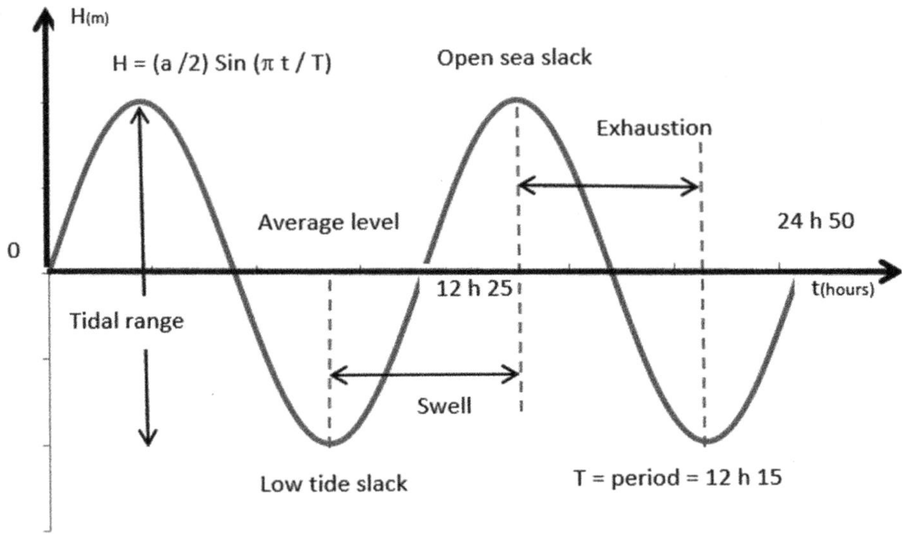

Figure 2.30. *Diagram of the semi-diurnal tidal cycle*

For example, the minimum pressure recorded in "Hugo", one of the most violent cyclones we know, was 918 h.Pa on September 15, 1989, at 18 h, before its arrival at Guadeloupe. This corresponds to a rise in the average sea level of 0.94 m (not counting the effects of the cyclonic wind).

The movement effects of wind on bodies of water are also well-known. The water, carried to the surface by a strong and constant wind, accumulates in the part of the water body below the wind, thus raising the level by several tens of centimeters. For example, in the Mediterranean on the coasts of the Gulf of Lion during the vicennial storm of December 16 and 17, 1997, in Narbonne, the water level rose to + 2.01 m NGF (*nivellement généralement de la France*, French general leveling benchmark), including about 1.45 m due to the effects of the wind.

In the study of marine infrastructures, storm surges and low levels are taken into account through the level reached by the sea; for example, we may want to implement dock coping out of water for given climatic events. Coastal defense works will have to be dimensioned according to the maximum water levels to avoid or limit flooding. In addition, daily variations in water levels can induce gradients in variations between high and low levels. For example, a relatively watertight dock, such as a combi wall or a diaphragm wall, will create a hydraulic gradient between the front and back sides. This hydraulic gradient between the front and the back will lead to an additional thrust, which will lead to stresses in the screen and to hydraulic effects due to the circulation of water in the soil which can lead to a risk of internal erosion of the soil near the wall.

Sea levels also condition the reference level of probes on all nautical charts. In France, the hydrographic zero or chart datum is close to the level of the lowest possible seas at high equinox tides (coeff. 120, HAT).

Thus, a probe marked 10 m indicates that there is 10 m of water remaining at the time of low tide. A probe marked 4.50 m indicates an uncovered seabed of 4.50 m above the lowest tide.

These coasts, which are called marine coasts (MCs), vary from one place to another on the coast and are locally correlated with the French general leveling benchmark (NGF), for which zero is generally higher and defined as the mean sea level at the tide gauge at Marseille.

Tides are generally accompanied by tidal currents, the intensity of which varies with the tidal coefficient and the speed of which decreases near the seabed. These local currents complement the large general ocean currents such as the Gulf Stream or the cold Labrador Current in the North Atlantic.

2.5.2.2. *Wind*

The rotation of the Earth and the resulting Coriolis forces cause air to flow over the Earth. The friction of this flow in contact with the surface of the ground or the sea gives rise to a surface boundary layer, a few tens of meters in height. It is only this layer that affects buildings and ships.

The flow being turbulent, the wind speed $V(x, y, z, t)$ can be considered as the superposition of the mean speed V (x, y, z), which can be assumed to be independent of time, and a fluctuation speed v (x, y, z, t), which can be assimilated to a stationary random process depending on space and time. Thus, the complete description of wind is defined, on the one hand, from that of the mean profile and, on the other hand, from that of the fluctuations.

On the basis of anemometer recordings, we can define the quasi-instantaneous velocity (effectively an average over a few seconds) and the average velocity (arithmetic average over 10 minutes of the quasi-instantaneous velocities). The quantification of the wind speed is fundamental in the determination of forces on ships, in particular according to their size and mass. This choice is generally different from that made for structures.

From the velocities, the action of the wind on the structures and constructions is basically expressed by a set of pressure fields (aerodynamic pressures); then, by simplification, by statically applied forces, whose effects on the studied structure are equivalent to the highest effects which are actually exerted.

The study of wind then involves the following steps:

– determination of "reference" velocities and pressures, by studying the local wind conditions (climatology);

– determination of the forces induced on structures and ships, according to the reference speeds, and then by taking into account the dimensions of the object exposed to the wind (usually called the frontal area);

– transmission of forces on maritime structures, such as a moored ship exposed to the wind that transmits forces to the dock by means of mooring lines, which is the function of a mooring plan;

– consideration of safety through wind speed (wind maps), and partial safety coefficients on wind forces.

2.5.2.3. *Sea states*

The previous two sections described tidal waves and wind. The latter, applied to a liquid surface, will generate natural sea states. Knowledge of sea conditions is essential for any coastal or harbor development project: (i) in climatology and for sedimentology studies (swell statistics over long periods, coastline stability, impact of developments, etc.); (ii) to define harbor agitation for studies related to the maneuverability of ships, their mooring and the comfort of harbor basins; (iii) to know the storm swells necessary for the design of port structures.

The sea is a liquid medium that is always in motion due to the effect of waves at different times and general currents of varying uniformity. It is the swell that is produced by the effect of wind on the surface of the water. At sea, when a wind starts to blow on a body of water that is initially at rest, small ripples appear due to the viscous friction of the air on the water. Their amplitude will increase to give waves of different directions, wavelengths and heights, depending on the regularity of the wind. This creates what is called a sea breeze. If the wind increases in intensity, the disordered agitation becomes structured and is transformed into a two-dimensional phenomenon that progresses in the direction of the wind. This is a cylindrical swell. The profile of the waves is strongly asymmetrical, and when the waves have reached a sufficiently high height, they reach the camber limit, which then leads to their breaking.

The swell is thus generated by the action of wind on the sea surface. It is then propagated from the storm site as a free oscillation whose characteristics change as it approaches the coast through various effects (refraction, diffraction, breaking, wave interactions, etc.) (Figure 2.31).

Figure 2.31. *Photograph of the swell (photo: course materials from the École des Mines d'Alès, Y. Stassen)*

The swell consists of waves that are more or less symmetrical and whose wave trains are irregular. It is therefore difficult to determine which representative measurements should be used to characterize the swell: maximum height, average height or characteristic value.

It is therefore necessary to use statistical models to calibrate models of random functions and probability laws. The characterization of a sea state is therefore generally a "spectrum", representative of all the components constituting the real swell. The characterization of waves themselves is defined by: (i) a height H or amplitude (peak to trough); (ii) a wavelength L and/or a time T (these two parameters are related by the celerity c = L/T); (iii) a direction of propagation (Figure 2.32).

When the swell propagates in variable depths, the wavelength, amplitude and direction undergo changes, unlike the time which remains an invariant: this is the phenomenon of refraction. When the swell approaches the coast, the waves undergo a rotation that renders them parallel to the base lines.

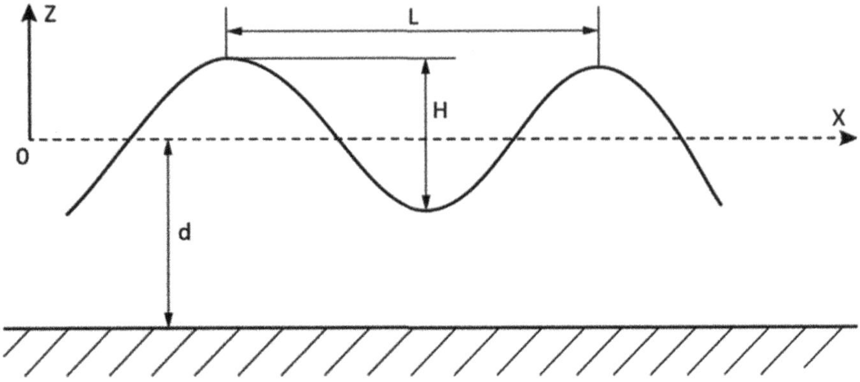

Figure 2.32. *Characteristic parameters of the swell*

As the depth decreases, there is a decrease in wavelength and a change in amplitude. The waves deform and the camber increases with the decrease in the depth. Beyond a camber limit, when the speed of the particles reaches the speed of the waves, the latter collapse and break in one of three ways, depending on the slope of the seabed (Figure 2.33).

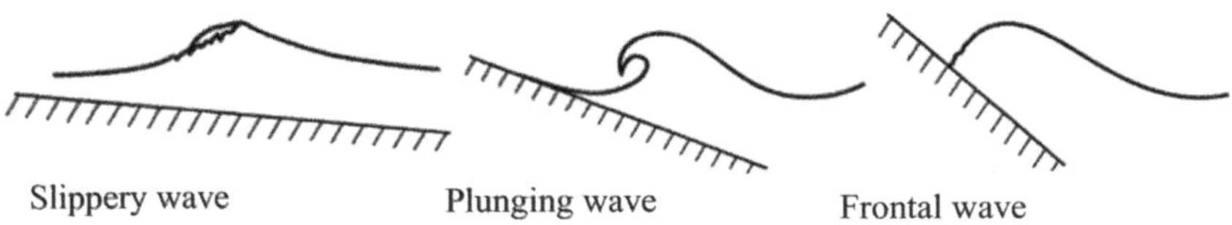

Figure 2.33. *Three types of breakers according to the slope of the seabed (source: Chapon (1966))*

When the swell reaches a vertical wall perpendicular to its direction of propagation, the swell reflects and creates a standing wave (fixed position of the nodes and antinodes) called a “rip tide”, the amplitude of which is double that of the incoming swell (Figure 2.34).

When the swell reflects off a sloping wall, the reflection is partial and can decrease to 60/70% of the total reflection for rockfill blocks arranged on a moderately sloped embankment.

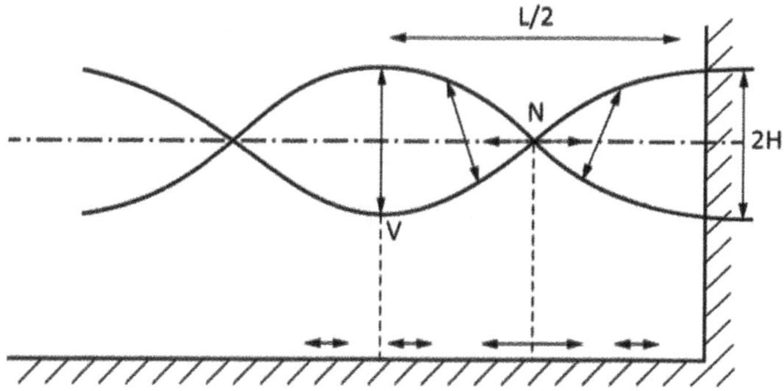

Figure 2.34. *Reflection of the swell on a vertical wall and the formation of the rip tide (source: marine engineering course materials (Chapon 1966))*

Finally, when the swell encounters a singular point such as a point, a cape, a pass, a dike, etc., it behaves like a swell emitter and impacts the amplitude but not the period or the wavelength. This phenomenon is called diffraction (Figure 2.35).

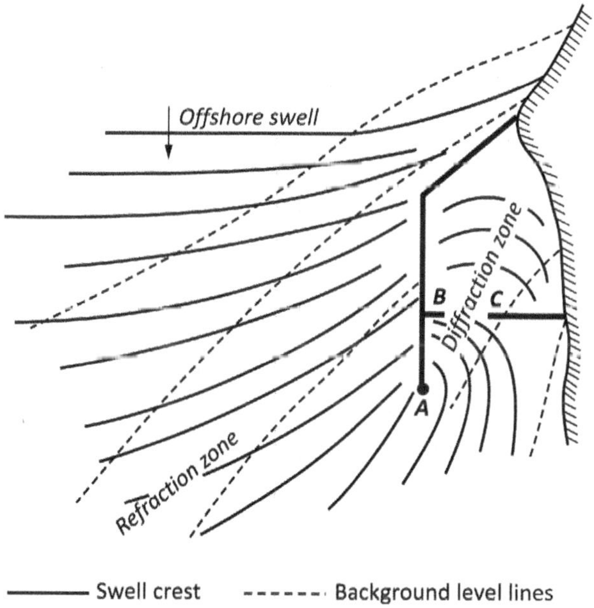

Figure 2.35. *Wave diffraction in the vicinity of a harbor breakwater (source: marine engineering course materials (Chapon 1966))*

Near the coast, the combined action of tides, currents and swell will shape the seabed and the shores. The swell has an erosive effect and tears off the particles, which the currents then transport.

The determination of hydrodynamic activity on the infrastructures is the subject of specific studies using the science of maritime hydraulics. These studies make it possible to determine the approach planes of the swell by taking into account the phenomena of diffraction and reflection at the edges of the body of water. From this, we deduce the agitation of the water body near the infrastructures and the forces applied to the structures. These studies call for sophisticated calculations that require powerful and specialized computing resources.

In summary, the establishment of a marine structure project requires, from the beginning of the project, knowledge of the environmental and oceanic conditions, such as: (i) water levels (tides, decreases, etc.); (ii) agitation in the area of the structure; (iii) winds and currents; and (iv) constraints relating to the respect of the environment (dredging limitations, species protection, etc.).

2.5.3. *Service actions*

The previous section highlighted the major ocean actions that have been identified and qualified. Many service constraints must also be considered in the design of a marine structure. These relate to: (i) project vessels; (ii) vessel loading and unloading operations; (iii) operating loads applied to docks and wharves; and (iv) equipment to be incorporated into the structure: mooring bollards, berthing fenders, rails and inserts for tools, bunkering or collection networks.

Knowledge of the project vessels is therefore essential. These are characterized by:

– type: pleasure, fishing or trade (passengers, solid or liquid bulk, containers, ferries, etc.). The type directly influences the equipment to be incorporated into the structures and their superstructures;

– geometrical characteristics: length, width, draft, displacement, frontal area exposed to wind and volume of hull exposed to swell and currents.

Similarly, the planned berthing and mooring arrangements (mooring plan) are fundamental data and guide the design choices (Table 2.4).

	Plan view	Cross-section
Mooring on the quay	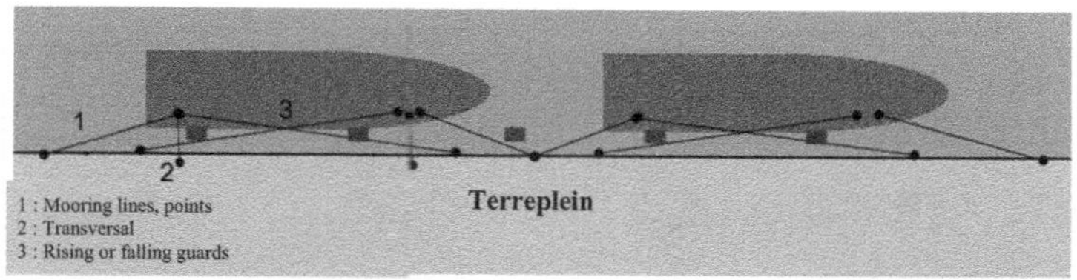	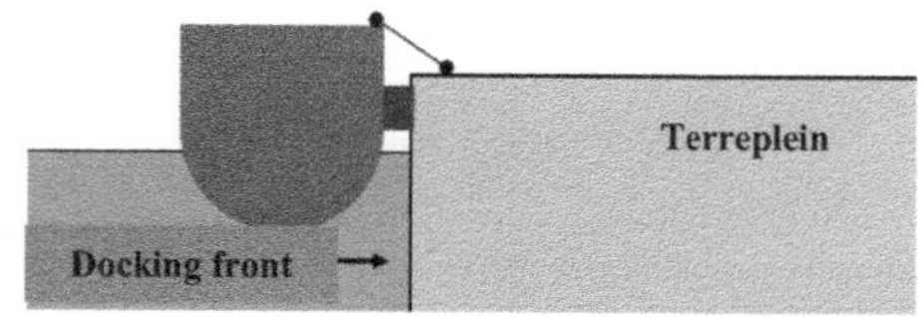
Mooring on a wharf	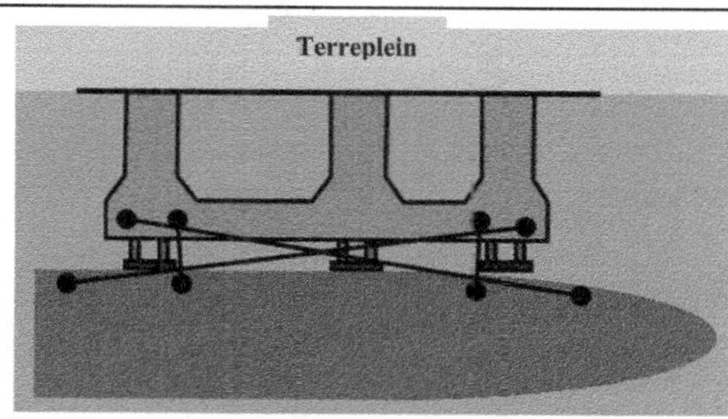	

	Plan view	Cross-section
Mooring on landing stage, jetty or breakwater	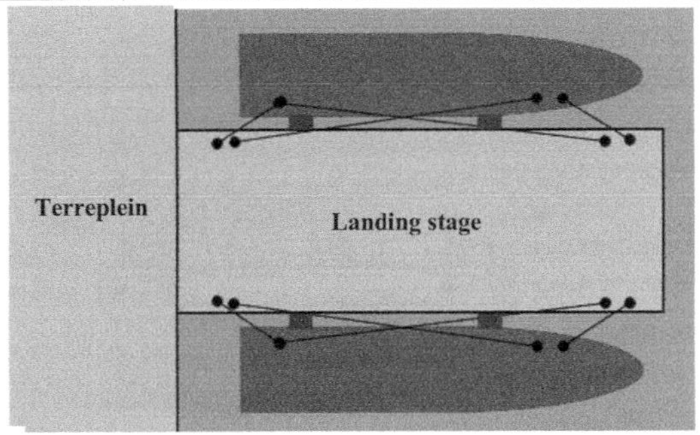	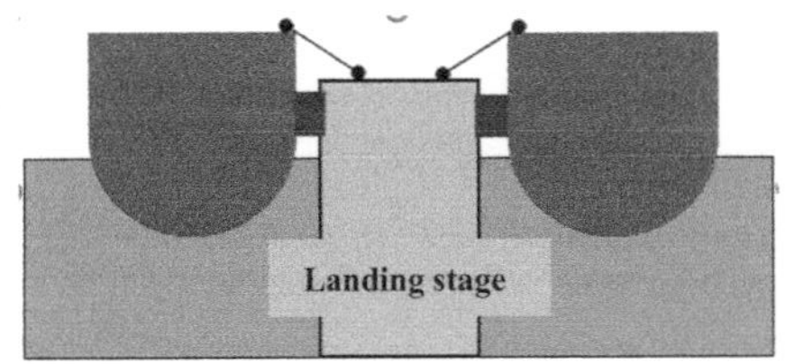
Anchorage on dolphins	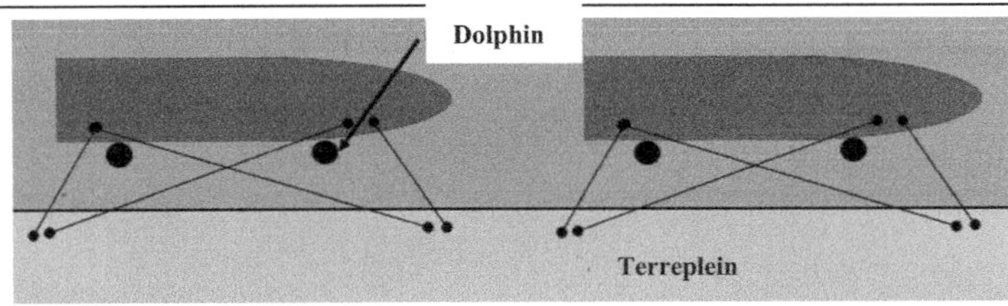	

Table 2.4. *Different mooring and berthing arrangements (source: marine works course materials – B. Sigros, J.-M. Cathala and J.-C. Souche)*

All these technical and functional elements influence the design of the structure and impose a choice of structure. In marine works, rustic and mechanically redundant structures are always chosen because the operating conditions are harsh. The infrastructures are often subjected to shocks and very aggressive conditions of use.

The projected works must also be adaptable to service evolutions during the projected period of use. They must be able to accept changes in equipment or different functionalities. Indeed, they have a minimum service life of 50 years (sometimes 100 years), whereas the evolution of ships and handling and load conditioning resources is much faster. It is enough to remember that, 50 years ago, ships were much smaller and that packages were moved by dockers (bags, crates) and by the ships' loading masts. Storage took place in sheds on the quayside!

The port infrastructure must include equipment essential to the operation of the port, without which the raison d'être of the maritime infrastructure does not exist:

– leisure: bollards, user access (stairs, gangways, catways, etc.);

– fishing: ice, connection with tidal sheds, handling equipment;

– bulk trade: loading/unloading machines (solid bulk) or canalizations (liquid bulk);

– container trade: gantries, mobile cranes, elevators, trailers;

– ferry trade: Ro-Ro station, inclined landing stage, support and mooring points;

– passenger trade: passenger flow studies, links with the maritime station, boarding bridge, etc.

All equipment therefore produces an equal number of service loads to be taken into account in the design of marine infrastructures: permanent loads, variable loads and load combinations.

2.5.4. *The usual design criteria*

The design of the works is the most difficult phase, as it involves:

– knowledge of all the possible solutions;

– assimilation of the particular constraints of the site (ground, sea, weather);

– meeting all the specified service constraints;

– finding the most appropriate economic response to the client's request, not only at delivery, but also throughout the operation of the structure.

The determination of the most economically advantageous solution must be based on a realistic construction scenario, adapted to the technical resources available in the vicinity, such as a usable dry dock, the availability of cranes or special pontoons and the availability of land.

The design of maritime structures must therefore be carried out in well-defined stages to determine the characteristics to be taken into account for the project:

– sea levels, ocean activity;

– geotechnical model: levels to be taken into account for the ground and the hydraulic water table, choice of characteristic values for the project;

– type of boats and geometrical characteristics: expected functionalities;

– safety margin, coastal seabed (depending on the ship and the agitation);

– berthing energy, mooring plan, mooring forces (depending on the vessel and site conditions);

– operating overloads and tooling constraints.

Table 2.5 provides criteria for selecting alternatives to consider or avoid based on the criteria listed.

2.5.5. *Geotechnical data*

As can be seen in Table 2.5, the geotechnical criterion is a determining factor in the choice of design for a marine structure. It is often an eliminatory criterion for some solutions.

On hard soils, weighted structures such as weighted walls and caissons play a supporting role and transmit loads to the surrounding soil through the shallow foundation that represents the base of the weighted structure. It is then necessary to verify the external stabilities of the weighted structure: sliding, overturning, punching and decompression of the foundation soil and finally large sliding. These external stabilities are totally dependent on the characteristics of the foundation soils and the embankments located behind the structure.

Determining criteria	From experience, we advise	From experience, we do not recommend
Soft, compressible soils	Pile docks	Weighted structures
Intermediate soils	Pile/shield docks	
Hard soils	Weighted structures. Works on piles only if anchoring by boring/drilling is economically possible	Screen docks due to the problem of embedding feet
Irregularity of "hard" soil levels	Prefabricated structures on a non-scourable distribution layer. Driven or drilled piles (difficult due to the criterion for stopping the driving on a case-by-case basis, pile by pile)	Creation of multi-level "stepped" pits should be avoided
Uneven soils (presence of erratic blocks)		Driven piles (except when accepting the risks from boring/drilling)
Tidal waters	Weighted docks: use of buoyancy as a means of lifting	Screens: treatment of joints to prevent material leakage
High operating loads (> 4t/m^2 approx.) or high point loads (rubber-tire cranes)	All types of docks, except high slab pile docks	High slab pile docks
Ships with profiled bows or overhanging superstructures	Spreading devices (important to prevent the ship from remaining on the dock in tidal waters)	

Determining criteria	From experience, we advise	From experience, we do not recommend
Multi-level docking in tidal waters	Screens, weighted structures or, more generally, continuous docking fronts	Pile docks without special docking structures
Risk of accidental docking		Sheet pile gabions
Significant docking loads on the quay	Distribution structure to support a high inertia edge-to-edge beam	Pile docks with load bearing by embedding the piles in slab beams
Earthquake	Pile docks: possibility of all types of docks if soil can be improved through the use of ballasted columns, for example	Retaining walls, especially if the soil is compressible
Work on land (dock to be excavated in the second phase)	Rigid wall (molded wall)	
Work in water		Rigid wall (except for temporary backfill to allow for work)
Networks on the quayside	Preferably in covered gutters out of the way of machinery traffic	Avoid networks under pile dock slabs, which are difficult to maintain and to upgrade

Table 2.5. *Criteria for the choice of technical solutions according to determining criteria (source: maritime works courses – B. Sigros, J.-M. Cathala and J.-C. Souche)*

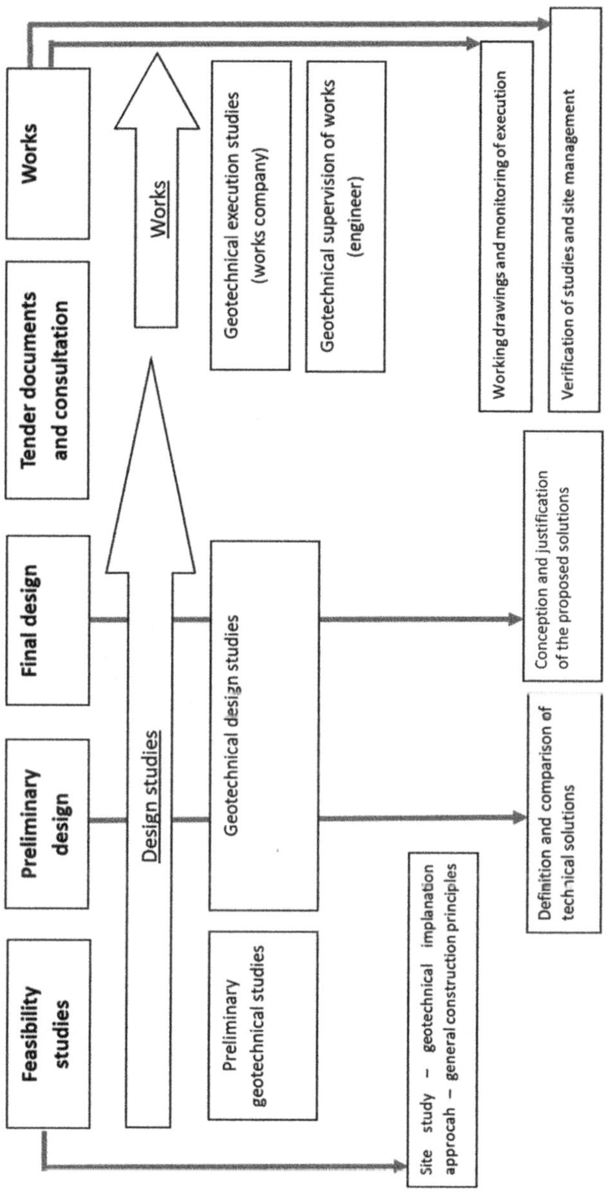

Figure 2.36. *Stages of project implementation and phasing of associated geotechnical investigations*

In intermediate soils of average characteristics, retaining walls are particularly suitable. The upper part of the retaining wall transmits the force to the lower part, which then begins to stress the abutted soil. The soil must therefore be sufficiently "flexible" to allow the screen to be driven into it, but sufficiently good so that the earth abutment allows the mechanical equilibrium to be reached, which guarantees the stability of the platform.

When the upper layers of the soil have poor characteristics, pile dock systems, based on deep foundations, are required. The piles work under vertical and horizontal actions. The conditions for embedding the piles at the foot and their connection with the upper reinforced concrete structure are very sensitive to the dimensioning of the structure. Pile docks are often associated with an underlying embankment which enables the levels between the docking front, where it is necessary to have the draft adapted to the ships and the platform, generally located 15–20 m higher, adjusted. The complete stability of this slope is a major factor in the overall operation of the dock.

In all the proposed cases, geotechnical data are fundamental. They must be determined as accurately as possible with regard to: (i) site geology and stratigraphy; (ii) soil identification; (iii) mechanical characteristics; (iv) hydraulic properties. These influence the chosen design, but the last of them also impacts the nature of the data required. Indeed, the geotechnical tests required to characterize the bearing capacity of a retaining wall or that of a surface foundation or a pile will not be the same. Consequently, a geotechnical study must be carried out in stages, in accordance with the phasing of the design studies, as detailed in Figure 2.36. In France, this phasing is the subject of a standard (NF P 94-500) that specifies the sequence of geotechnical missions (Figure 2.36).

The particularity of reconnaissance for maritime works comes from the development of aquatic sites which are difficult to access and involve the presence of water. Of course, this poses logistical problems, relating the site conditions, but also geotechnical problems in determining good conditions to carry out the samples and tests.

Acquiring geotechnical data is essential for the proper design of marine structures. However, it is expensive and the investigations must be carried out by a highly competent geotechnician with the right equipment.

2.5.6. *Data on construction materials*

2.5.6.1. *Effects of the sea on steel and concrete*

Sea water has a composition and chemical properties that make it aggressive towards construction materials, whether it is concrete or the steel that makes up all types of pipes and metal sections (sheet piling, I-beams/H-beams, etc.).

The salinity of sea water, which is the total weight of dissolved mineral salts (NaCl at nearly 80%) has a mean value of 35 g and varies from one sea to another (43 g in the Red Sea, 38 g in the Mediterranean Sea, 32 g in the North Atlantic Sea and 10 g in the Baltic Sea).

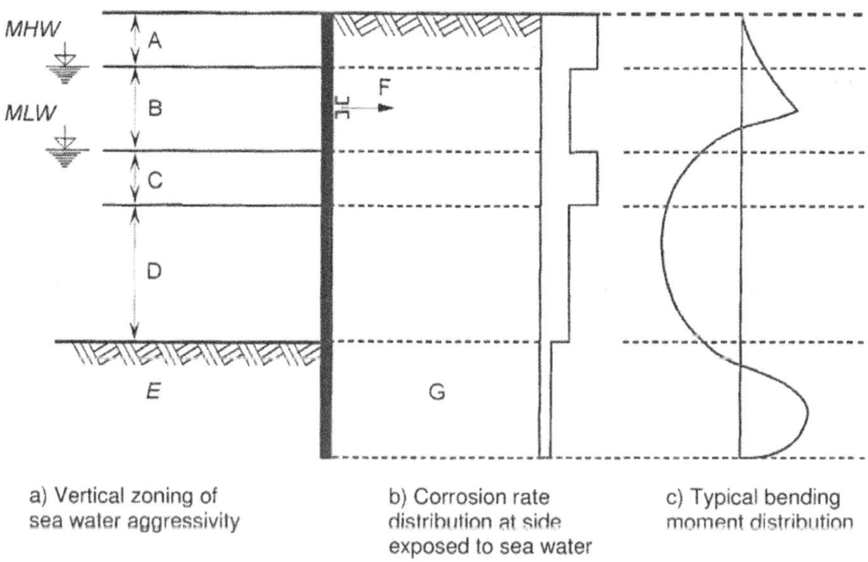

A	Zone of high attack (splash zone)	B	Intertidal zone
C	Zone of high attack (low water zone)	D	Permanent immersion zone
E	Buried zone (water side)	F	Anchor
G	Buried zone (soil side)		
MHW	Mean high water	MLW	Mean low water

Figure 2.37. *Identification of high corrosion zones based on water levels (source: Eurocode 3 part 5)*

The action of sea water on structural steel is due to the creation of a galvanic battery, where salt water in a liquid state or in the form of spray is the electrolyte. In practice, these batteries are formed when there are several metals close to each other (such as steels of different grades) or when there is heterogeneity on the surface of the steel. This creates anodic and cathodic zones. Each time the steel is in the anode position, it will corrode.

In the case of steel, corrosion results in the formation of rust, which is composed of oxides that only form in the presence of oxygen and water at normal temperature. The electrolyte is water, with varying degrees of conductivity. The more this water is oxygenated, with a high temperature or salinity, the more significant its corrosive power. Eurocode 3 part 5 specifies the aggressiveness of the environment for metallic structures depending on the degree of exposure to sea water and air (Figure 2.37).

These considerations lead to either sacrificing thickness to corrosion, by projecting to the end of the structure's life, or use of a anticorrosion agent, which can effect satisfactory protection for the infrastructure's duration of use.

The action of sea water on the hydraulic binders of concrete and on the reinforcements of reinforced or prestressed concrete structures is due to the chemical reaction of lime and magnesium sulfates on the components of cement, free lime, portlandite and alumina in particular.

The first effect is an *attack on the concrete* by sea water, which can take different forms:

– lixiviation of the portlandite under the hydraulic effects of the sea (air/water alternation in tidal zones, presence of currents, effects of the swell on the concrete works, etc.), which will increase the porosity of the concrete;

– attack on the cement matrix by sodium chloride; in the presence of water, chloride ions penetrate the porous medium of the concrete. On the one hand, this results in the consumption of portlandite by the formation of soluble calcium chloride ($CaCl_2$). On the other hand, the tricalcium aluminate C_3A and the hydrated aluminates react with the calcium chlorides to form calcium mono-chloroaluminate ($C_3A.CaCl_2.10H_2O$), which can undergo a sulfate attack and finally form ettringite. This compound causes a significant swelling and a bursting of the concrete with exposure of the reinforced concrete and corrosion, then loss of mechanical resistance of the reinforced concrete;

– attack on the cement matrix by magnesium chlorides, which exert a dissolving action on the lime and then lead to the formation of brucite ($Mg(OH)_2$) which can also lead to the formation of calcium mono-chloroaluminate, as before;

– attack on the cement matrix by magnesium sulfates ($MgSO_4$). This salt is very aggressive and has multiple effects: (i) dissolution of Ca^{++} ions from the hydrates; (ii) exchange with Mg^{++} ions from the sea water. These reactions lead to the precipitation of brucite ($Mg(OH)_2$) and to the progressive transformation of C–S–H into C–M–S–H and M–S–H without binding properties, as well as to the formation of gypsum. As before, the reaction of the sulfate with the aluminates of the binder can lead to the formation of ettringite, which may be expansive. The lixiviation of C–S–H and the expansion of ettringite cause an increase in porosity and cracking that enhances the weathering processes of the cement matrix and the reinforced concrete;

– the action of carbon dioxide from sea water and air in tidal areas. Calcium hydroxide and C–S–H combine with carbon dioxide to produce calcium carbonate ($CaCO_3$) which precipitates as calcite. This results in a drop in the pH of the concrete, which then makes corrosion of the reinforcement possible.

The second effect is the *corrosion of reinforced concrete* by sea water. Under the effect of its lixiviation, carbonation, or the attacks of chlorides and sulfates, the concrete, which is initially a protective medium of the steel where the reaction of corrosion is inhibited[13], loses this fundamental property and the corrosion of the reinforcements is favored. This corrosion leads to the degradation of the steels (loss of sections) and generates expansive products that cause the concrete cover to crack, which facilitates the penetration of the aggressive agents and accelerates the general process of deterioration.

2.5.6.2. *Protection of steel structures*

The first action to be taken is to correctly select the grade of steel to be used in the structure. In general, this is determined according to the forces to be absorbed and the stress level reached in the sections.

To prevent the loss of metal thickness due to corrosion from being too great, steel structures are usually protected from corrosion by various techniques:

– anticorrosion protection by painting: this is the usual method for protecting metal foundation piles from corrosion. The paint system is chosen from the list of systems certified by the ACQPA[14], which offers guarantees of the longevity (>15 years) and color of the paints used. The paints are generally applied in the factory under controlled temperature and hygrometric conditions and after preparing

13 The hydration of the cement produces a basic interstitial solution with a high pH of about 12–13. Under these conditions, the reinforcements are protected because the chemical reaction of corrosion cannot develop for pH > 9.

14 *Association pour la certification et la qualification en peinture anticorrosion* (French Association for Certification and Qualification in Anticorrosion Painting).

the supports by shot blasting. The application of a well-chosen and well-applied anticorrosion system will enable a closed physical barrier between the steel and the natural environment to be created;

– galvanizing corrosion protection: the principle consists of covering the steel with a thin layer of zinc, which will protect the steel from corrosion. Indeed, oxidation of zinc is slow compared to that of steel. Zinc also provides sacrificial protection in the event of damage or coating discontinuity, due to the stacking effect between steel (cathode) and zinc (anode)[15]. Structures usually protected by galvanizing are chains or structural steel elements of reasonable size, which can then be treated by immersion in a bath;

– cathodic corrosion protection: for piles, anodes made of a metal more electronegative than steel are placed near the metal element that constitutes the cathode of the resulting galvanic pile; for sheet piling, an electrical source is properly connected so that the steel of the marine structure becomes the cathode of the system, which makes its corrosion impossible. This is called impressed current cathodic protection.

2.5.6.3. *Specifications for the concrete material*

By way of example, for the concrete material, the standard NF EN 206/CN 2014 (AFNOR 2014) translates a prescriptive approach based on an obligation of resources relating to the resistance and the durability of concrete. The requirements of this standard are based on the aggressiveness of the environment of the structure and its duration of use.

The project service life is defined by Eurocode 0 and corresponds to the service life of the structure in normal operation and with routine maintenance. Project service lives are typically 50 years for buildings and 100 years for civil engineering structures.

Sometimes, for marine structures subject to an ATO[16], the maximum duration of which is 30 years under French law, the duration of use may also be 30 years. After this period, state services could request that the structure be dismantled and packed away.

Exposure classes reflect the type of external attack that a structure may undergo during its service life and the increasing level of aggressiveness for each attack considered. They are also based on three types of concrete: (i) unreinforced

15 Be aware of the effect of the release of Zn, Al or other metal ions during normal operation of the galvanic cell and its effect on the environment!

16 Authorization for temporary occupation of the public maritime domain.

concrete (UC); (ii) reinforced concrete (RC); and (iii) prestressed concrete (PC). The exposure classes are defined in Table 2.6.

	Exposure classes	Comments
	X0	No risk of corrosion or attack
Risk of corrosion	XC1–XC4	Carbonation-induced corrosion
	XS1–XS3	Corrosion induced by chlorides in sea water
	XD1–XD3	Corrosion induced by chlorides other than marine ones
Attacks	XA1–XA4	Chemical attacks
	XF1–XF4	Freeze–thaw attacks with or without de-icing agent
Abrasion	XM1–XM3	Moderate to extreme abrasion from vehicle traffic

Table 2.6. *Main exposure classes of concrete*

For marine structures, the most common classes are XC and XS. They are listed in Table 2.7 by type of structure.

For parts of structures that are "permanently immersed"[17], the class will be XS2.

For the classes relating to chemical attacks, the choice between XA1, XA2 or XA3 will be made according to the level of aggressiveness of the polluted sea water and/or ground water that may be in contact with the structure.

The abrasion classes are to be chosen between XM1, XM2 and XM3, according to the level of aggressiveness of machine or ship traffic scraping the docks. Moreover, the risk of abrasion is to be assessed according to the level of aggressiveness of the swell and the currents circulating in and around the structure, the friction of mooring lines, propeller jets and the presence of sediments, floating anchors or abrasive materials which can be put in suspension, etc.

The prevention of the phenomenon relies on the resistance to abrasion of the concrete material or even the installation of metal parts locally. In the case of reinforced concrete, this risk is expressed by the choice of the abrasion class XM1, XM2 or XM3 and the prevention of abrasion may also rely on a sacrificial thickness, coating as provided for by the NF EN 1992-1-1 standard.

17 It can be ensured that they are located, for the entire period of intended use of the structure, below the spring tide low-water mark, except in the case of wet docks where this limit is determined by the mode of operation of the body of water.

Type of structures	Classes	UC–RC–PC
External structures (protection of the harbor from the effects of the swell)		
Breakwaters	X0–XA1	Unreinforced concrete (UC) blocks
Dikes, jetties, breakwaters	Permanently immersed parts: XC1–XS2 Aerial parts/tidal zone: XC4–XS3	Reinforced concrete (RC) caissons on ballast/dock of RC blocks
Heavy pontoons Breakwaters	Anchors: XC1–XS2 Floating pontoon: XC4–XS3	Concrete pontoons (RC) Concrete anchors (RC)
Port structures with port service functionalities		
Wharves	Permanently immersed parts: XC1–XS2 Aerial parts/tidal zone: XC4–XS3	RC deck/massive RC caisson pillars
"Danish" wharves (wharves on piles), landing stages, wharves		RC or PC deck on steel piles or reinforced concrete
Screen docks or caisson docks	Permanently immersed parts: XC1–XS2 Aerial parts/tidal zone: XC4–XS3 Fully buried reinforced foundations: XC2–XS1	Concrete caissons Molded wall-type RC screen
Pontoons	Anchors: XC1–XS2 Floating pontoon: XC4–XS3	RC pontoons RC anchors

Types of structures	Classes	UC–RC–PC
Locks/gates	Sinkhole: XC2–XS2 Bays: XC4–XS3	RC infrastructure
Port structures dedicated to the maintenance of ships		
Dry construction docks, dry slipways, slipways	Permanently immersed parts: XC1–XS2 Aerial parts/tidal zone: XC4–XS3	UC, RC RC, PC
Self-lifting platforms/boat elevators		RC infrastructure
Open sea structures of various functionalities other than protection		
Anchors, weights	Permanently immersed parts: XC1–XS2	RC
Artificial reefs		UC, RC
Pipelines and outfalls in the sea	Permanently immersed parts: XC1–XS2 Aerial parts/tidal zone: XC4–XS3	RC
Offshore structures (for the record)		RC infrastructure

Table 2.7. *Main exposure classes for concrete structures used in marine areas (source: CIMbéton documents)*

For each of the exposure classes, the EN 206 standard fixes the specific limit values applicable for the composition and the properties of concrete that must be respected: minimum binder dosage, maximum water/binder ratio, type and class of authorized constituents, class of minimum compressive strength of concrete, etc. Criteria must also be respected for the choice of constituents and for the conditions of implementation with respect to the alkali-aggregate reaction (AAR) and with respect to the internal sulfate attack (ISA).

The result of these provisions is a definition of a concrete that takes the following form (Figure 2.38).

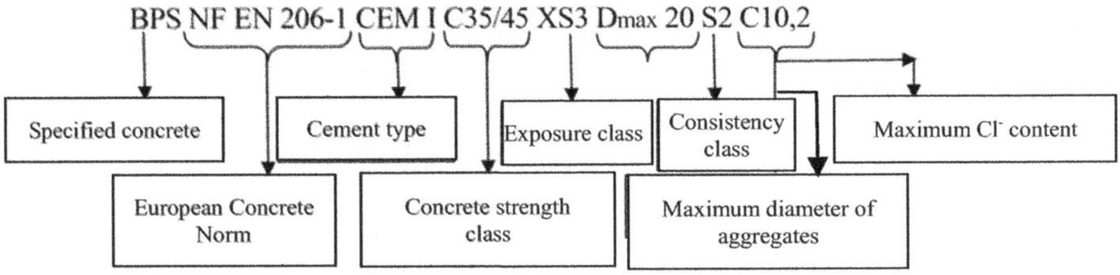

Figure 2.38. *Typical identification of a concrete for the ready-mix concrete plant*

This approach does not take into account certain parameters relevant to durability. It focuses on the composition of the binder and reduces the use of additions. Such an approach does not allow for the qualification of certain concrete compositions (compound concretes, for example), which are technically interesting and equal in cost. It sometimes translates contradictory requirements and does not facilitate either innovation or the use of recycled materials.

The standard makes it possible to correct these deficiencies by calling upon the notion of a performance approach for durability, which is an innovative, global and predictive approach to the durability of reinforced concrete structures. It is based on: (i) the concept of durability indicators; (ii) a scientific approach that uses calibrated models relative to the different degradations taken into account; and (iii) local concrete formulas, if possible, for economic purposes.

The performance approach does not impose the composition of the concrete, the choice of the binder or the aggregates[18]. The justification of the concrete formula requires durability tests on the concrete (performance tests and/or durability indicators). The thicknesses of the mixes can therefore be modulated according to the actual durability properties of the concrete (Figure 2.39).

---

18 The constituents must conform with the NF EN 206 standard that specifies safeguards for the composition and formulation of concretes.

Contrary to the standard approach, we do not have an obligation of resources but of results.

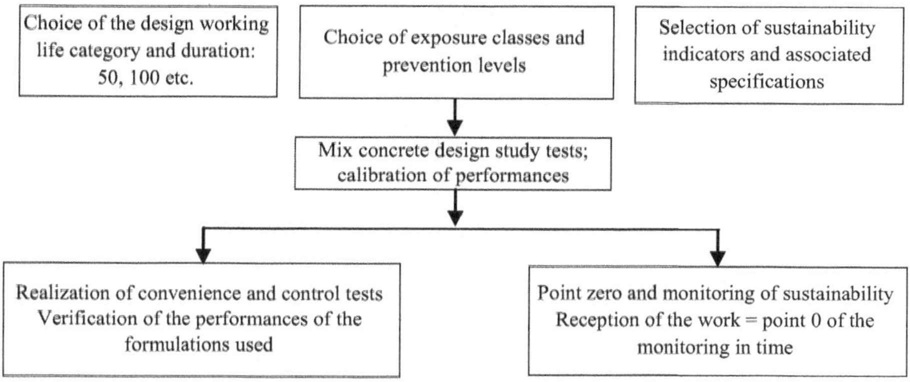

Figure 2.39. *Synoptic of the approach of performance formulation of concretes (NF EN 206)*

2.5.7. *Synthesis*

At this stage of the process, we have not considered any given eco-design of marine structures. However, the process for the civil engineer is complicated enough. Figure 2.40 shows the process to which they are subjected and to which they must respond for many technical points.

On the basis of the program and the needs identified by the client, the civil engineer will compile the existing data, manage the acquisition of additional data and propose a feasible design from all points of view, which will be the subject of a preliminary design document.

An environmental engineer will then carry out the regulatory environmental studies that will specify the feasibility and the some of the major constraints to be respected for the structure in service, as well as during the works.

The civil engineer will then integrate these environmental constraints in their final project, which will be used as a basis for the elaboration of the work contracts of the contractors.

Contractors will carry out the work until the acceptance and transfer of ownership to the owner and the start of operations. The work will be managed by the civil engineer, who is responsible for quality control and for ensuring the work is on time and on budget.

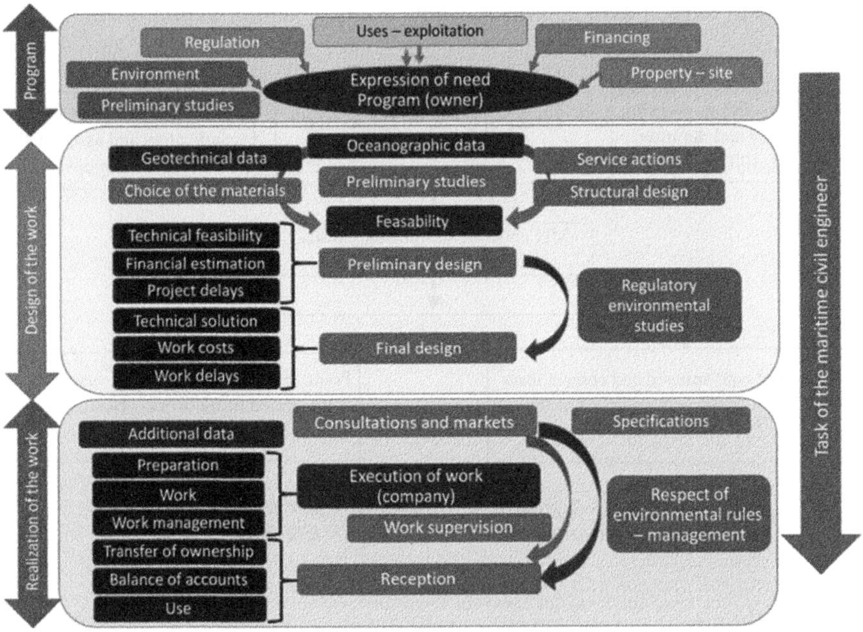

Figure 2.40. *Synoptic outline of the tasks and missions of the marine civil engineer*

In this typical approach, the task of the civil engineer is already dense and complex. They must call upon all their skills and all the experience acquired during previous projects. They must show pragmatism and common sense to find the right design of structures that meets the needs of the operator, integrates the numerous constraints and data, offers an economic optimum and is justified by sufficient calculations that guarantee the requirements of mechanical resistance, such as durability. Their responsibility is engaged on these crucial points.

They disregard the mechanisms and methods that might be needed to understand the environmental aspect, perceiving it instead as nothing more than the sum of additional constraints. However, it would be wrong to believe that they are not concerned nor ready to integrate ecological concepts. The art lies in a method that we propose to set out in Chapter 3.

On the contrary, the integration of environmental concepts into the design of a structure gives an additional meaning to the engineer's activity. It gives them the possibility of being part of the virtuous process of sustainable development. This is extremely rewarding for a construction professional.

3

Eco-design of Marine Infrastructures

As Aronson *et al.* (2016) remind us, "Engineers and ecologists must work together and learn from each other if our work is to generate significant societal benefits." Mitsch (2014) asks, "When will ecologists learn engineering and engineers learn ecology?" The current lack of integration between these two professions is one reason why the creation of new infrastructures that successfully integrate ecological and engineering concerns is so rare. Another reason is that marine infrastructure eco-design is currently aimed at professionals who are committed to a voluntary approach because they are aware of the benefits it brings.

In our experience, although some educational programs are developing courses and research programs on marine ecology, currently most civil engineers are not really aware of, or sensitive to, ecological issues when tasked with developing projects, or when they are specifically asked to "build with nature". This is not due to a lack of interest in ecological issues or lack of appropriate training, but rather to the current philosophy of planning in natural areas. In parallel, due to a great ignorance of construction processes, "ecologists" can adopt principled positions that are irreconcilable with the development project, or at least with a pragmatic vision of it.

As a result, we deplore the fact that technical, economic or social concerns still have a clear priority over environmental (especially biological) concerns in the design of infrastructures. These are only seen as a set of additional constraints that are imposed on everyone, often to the detriment of the economic dimension of the project.

However, we believe that it is possible and realistic to chart a new course in the design of maritime structures: eco-design. This is the objective of this book, as Richard Spieler explains in Box 3.1.

"My group, with colleagues, has been extensively involved in research with artificial reefs designed for and used in mitigation. However, we have barely made a dent in the information required to fully understand the structural requirements of effective marine eco-design. The complex issues of artificial structures interacting with the biota at any given site are daunting. Yet with ever-increasing human populations and the concomitant shoreline development this knowledge is an absolute necessity to maintain the quality of life we derive from marine resources. Not only is the basic knowledge required, such knowledge must also be effectively put to use. This, in turn, requires a sea-change in the thinking of resource managers, engineers and biologists. These groups have differing discipline-related goals and currently often act separately and occasionally at cross-purposes in relation to marine eco-design.

This book will provide a baseline inventory of past and current projects incorporating eco-design and highlight research needs for the future. It will also highlight the current problems with myopic discipline-related approaches to mitigation and marine construction."

Box 3.1. *Why a book on the eco-design of marine structures? (source: Richard Spieler[1])*

3.1. The evolution of research work towards the eco-design of marine structures

Marine eco-design is the result of an intellectual maturation in the relationship between development and colonization, which it is important to briefly retrace here.

Outside of Japan (a country that has been a forerunner in research on the creation of structures that support biodiversity), many researchers, particularly through work on artificial habitats, have studied the effects of substrates on colonization since the 1970s. It is therefore possible to design structures necessary for human activities, such as breakwaters, and favorable to the installation of biodiversity (Table 3.1).

The influence of different biophysical environmental factors has been studied in order to better understand how marine ecosystems develop (Persoone 1971; Hartog 1972; ZoBell 1972; Nickels *et al.* 1981; Connell 1985; Relini 1993; Bourget *et al.* 1994; Harmelin-Vivien *et al.* 1995).

Gradually, in the 1990s and early 2000s, authors also became interested in the effect of "conventional" artificial concrete infrastructures on ecosystems: ports, breakwaters, offshore platforms, etc. (Fletcher and Callow 1992; Connell 2001;

1 Professor of Oceanography Emeritus, Co-Dean of the Oceanic Center at Nova Southeastern University, Fort Lauderdale, Florida, USA.

Glasby *et al.* 2007). The results of several of the most significant studies, conducted in Australia, the United Kingdom and Italy, were surprising (Airoldi and Bulleri 2011). They suggested that not only were certain immersed coastal structures built from a concrete source of ecological distortion in relation to natural environments (different varieties of species with the rocky natural environment, low colonization diversity, potential pollution from the materials used, etc.), but that they were also vectors for the development of non-native, ubiquitous invasive species (Hillier *et al.* 1999; Bulleri and Chapman 2004).

For others, the positive impacts of infrastructures constructed from quarried boulders were interesting and linked to the diversity of forms used, introducing a substrate beneficial to local species, to the ecological functions created and thus to biodiversity (Ruitton 1999).

This double impetus from advancing knowledge in eco-engineering and the revelations about the occasionally negative ecological impacts of classic concrete infrastructures has led to an increasing number of publications in the last decade on the improvements of the impacts of these structures on the marine environment. The marriage between the "technical" design of a project and its optimization, with respect to the positive effects for ecosystems, is clearly developing.

Without directly linking to the positive effects for infrastructure, other authors continued to focus on shapes (designs) of structures, without substantially increasing costs, in order to develop shelters favorable for colonization (Li *et al.* 2005; Moschella *et al.* 2005; Jackson *et al.* 2008; Martins *et al.* 2010).

The trend that seems to be emerging now is one of dual benefits: how to ensure better sustainability and efficiency of structures while developing ecosystems (Borsje *et al.* 2011; Firth *et al.* 2020). This is the point Louise Firth makes in Box 3.2.

"With the increasing trend of human population increase and human impacts on the environment, governments all over the world are striving for sustainable development that works with nature rather than against it. Whilst a wealth of terms exist, such as nature-based solutions, ecological engineering and reconciliation ecology, they all effectively have the same goal – to make space for nature in a human-dominated world. This practice has long existed on land but is a much more recent field in the marine environment. In recent years there has been a surge of research in this field globally, with the emergence of a 'tool box' for managers to access when considering eco-friendly designs in marine and coastal environments. The majority of this research has been published in a wide range of scientific, engineering and landscape architecture journals, making it difficult for the practitioner to access. This book represents the collation of much of this work and provides a one-stop shop for all aspects of marine eco-design.

The EU-FP7 funded THESEUS Project (Innovative Technologies for Safer European Coasts in a Changing Climate) led to the testing of a range of different eco-design applications in the UK, including the development of the prototype precast habitat enhancement unit called the BIOBLOCK and the testing of drill-cored artificial rock pools in an intertidal breakwater. Both of these techniques have since been tested in other parts of the world in an effort to test their efficacy under a range of different environmental scenarios.

This has inspired the further wider testing of a range of different experimental techniques. As much of this research has been conducted on small spatial and temporal scales, it is advocated that the designs described in this book be repeated elsewhere and critically appraised for their efficacy. This will provide a stronger evidence base for this kind of research. This is particularly important as managers and planners are looking to the community for advice on how to incorporate ecological concepts into new and future designs."

Box 3.2. *Why introduce environmental concerns into design? (source: Louise Firth*[2]*)*

NOTE.– Many publications continue to denounce the negative effects of coastal infrastructures (invasive species, destruction of the seabed, toxic materials, etc.). We trace here the developments in studies on the role of infrastructure in the environment.

By way of example, Iwanami *et al.* (2002), Maruya *et al.* (2003) and La Marca *et al.* (2014) have demonstrated how concretions (fouling) on breakwater-type structures can extend the life of concrete, creating a co-benefit called "bioprotection".

Kawabata *et al.* (2012) demonstrated that the migration of chloride ions (one of the main factors in the aging phenomenon of marine concrete) was considerably slowed down by colonized by oysters or limestone worms, a physical barrier and thus a natural insulator for the structure. Firth *et al.* (2013) demonstrated the positive effects of colonization favored by the concept of eco-design for coastal defense structures intended for wave damping or beach erosion control: algae grow and dampen the effect of the waves; those that are removed add solid material that improves sand fixation on beaches.

Similarly, Coombes *et al.* (2015) highlighted the "eco-technological" (or eco-engineering) opportunities for the development of reworked concrete, as a biodiversity support for the fight against global warming and its effects: better resistance of structures to recurrent extreme climatic events, absorption of CO_2,

2 Senior Lecturer in Marine Ecology, School of Biological and Marine Sciences, Faculty of Science and Engineering, University of Plymouth.

filtration and recycling generated by developed ecosystems, improvement of the resilience of the natural environment, etc.

It is therefore possible to design structures that are necessary for human activities, such as breakwaters, and favorable to the establishment of biodiversity (Table 3.1).

However, eco-design is still struggling to become widespread because, in addition to its novelty and the lack of dissemination of good practices, it is not a "land" vision but a marine one, linked to the three dimensions of this territory.

An article published by Denis Lacroix in 2011, "Un nouveau territoire de développement : la mer côtière ou le développement en4D"[3], mentions this spatial specificity of the underwater world. It was presented as a catalyst for innovation that is still too extreme, in our view, for the "flat earthers" (Figure 3.1).

Indeed, when constructing a building, it is not obvious to consider the natural colonization of the facades, as it is easy to see that, even after decades, concrete walls are not covered with biota. In the sea, it is quite the opposite, with each mm^2 covered, at best, with a biofilm (made of cyanobacteria, then diatoms, only a few hours after immersion) and, at worst, with an assembly of a more complex fauna and flora.

The idea of eco-design for marine structures is, in fact, an ethical responsibility of "marine" planners, as opposed to that of land planners and architects, who are not subject to the same biological conditions of a "fertile" aquatic environment.

Moreover, the creativity of the architect will intervene for the emergent "habitable" areas of a marina or a port visible by humans, but never to orchestrate the conception (the design) and then the construction of the immersed areas of these structures, leaving the civil engineers to carefully calculate and design "purely" technical, functional and generally unattractive structures.

No doubt linked to an empirical approach to the sea, it is in the East, specifically in Japan, where concrete constructions based on more traditional principles can be found. Tanoue (2015) speaks of "ICZM[4] applied to the construction of a technical project". This idea is based on the theories of Yanagi, from the 1990s, which introduced the idea of "a contract with nature at sea to ensure our survival on a planet with limited resources".

3 This article received the first prize from the *Conservatoire national des arts et métiers* (CNAM) in 2010.

4 Integrated coastal zone management: concept and methods aiming at developing systemic, or integrated, projects on coastal zones, essentially multi-stakeholder and multi-purpose.

Coastal infrastructure research	Scientific reference literature	Phases – period	Effects on biodiversity
Natural colonization of artificial substrates	Zobell 1972; Relini 1993	1970–1990	No study
Risk factors related to colonization Concrete toxicity, disruption of landscape connectivity, indirect impacts (carbon footprint, sand and rock extraction, etc.) Invasive species, modification of natural dispersion patterns	Hillier *et al.* 1999; Wilding and Sayer 2002; Bulleri and Chapman 2004; Bulleri *et al.* 2005; Moschella *et al.* 2005; Glasby *et al.* 2007; Jackson *et al.* 2008; Bulleri and Chapman 2010; Terlizi and Fairmali 2010; Airoldi and Bulleri 2011; Dafforn *et al.* 2012; Jacob *et al.* 2016; Firth *et al.* 2016	Since the 2000s	Negative
Development of biodiversity (materials, roughness, surface treatment, holes, cracks)	Wilding and Sayer 2002; Lee *et al.* 2008; Martins *et al.* 2010; Chapman and Underwood 2011; Coombes *et al.* 2011; Pioch *et al.* 2015; Evans *et al.* 2016; Souche *et al.* 2016	Since the 2000s	Positive
Biofouling protection	Moschella *et al.* 2005; Coombes *et al.* 2013; La Marca *et al.* 2014; Perkol-Finkel and Sella 2014; Coombes *et al.* 2015	Since the end of the 2000s	Positive sustainability of infrastructures improved
Integration of ecological knowledge into coastal infrastructure, co-benefits => eco-design	Lacroix and Pioch 2011; Firth *et al.* 2014; Phan *et al.* 2014; Coombes *et al.* 2015; Dafforn *et al.* 2015; Sella and Perkol-Finkel 2015; Patranella *et al.* 2016; Coombes *et al.* 2017; Perkol-Finkel *et al.* 2018; Pioch *et al.* 2018; Salgues *et al.* 2020	Since 2010	Positive specific ecological goals related to improved sustainability

Table 3.1. *Evolution of research themes on the relationship between coastal infrastructure and marine ecosystems*

Figure 3.1. *Photomontage of an eco-designed offshore wind turbine foundation (Lacroix and Pioch 2011)*

Both of these theoretical and philosophical concepts are now consolidated within the eco-design approach, which allows projects to be designed technically while simultaneously combining technical and ecological concerns and without generating additional costs in the long term.

Eco-design therefore has definite advantages:

– the colonization of structures by organisms can provide ecological added value that improves the technical performance of a structure, as co-benefits. For example, the bioprotection of a dock by mussels can protect concretes from environmental aggression (penetration of chloride ions in particular) (Figure 3.2). The development of a biofilm on the surface can reduce the permeability of concrete (sealing of cracks and/or surface porosity);

– eco-design enables the integration of the development into its environment to be maximized by also promoting the construction of bioinspired and aesthetically pleasing, from an underwater point of view of course, structures (biomimetic immersed natural landscape structures);

– eco-design also benefits the project authorization process by reducing the risk of appeal due to a failure to take the environment into account. In addition, it encourages interaction with all project stakeholders from the outset.

This approach is integrated, "proactive" and not reactive on environmental issues. However, let us not forget that any new project destroys the natural environment and even an eco-design approach will lead to ecological losses, certainly reduced or partially compensated for, but, above all, avoided to the maximum.

Figure 3.2. *Presence of mussels (*Mytilus galloprovincialis*) and barnacles (*Chthamalus*) on the surface of the concrete, lower exposed zone of the coping H-beam of the dock, offering protection against physical and chemical effects (UV, chop, chlorides); port of Sète, south of France (photo: S. Pioch)*

3.2. The modernized approach to project management

Although it is well-established that it is necessary to take into account the environmental impacts of marine development, the implementation of these considerations is generally perceived by some as restrictive, complicated, difficult and sometimes utopian. It is often in opposition to project cost control, its technical feasibility and mindfulness of project milestones.

Indeed, if the project is sufficiently advanced to reach a level of mature technical and economic feasibility (the preliminary project stage, for example) without the integration of environmental considerations, then taking these considerations into account, when required, will lead to complicated, if not impossible, re-evaluations for the designer.

This often leads to environmentally nonoptimal solutions, missed opportunities and the definition of compensatory measures whose cost is real, sometimes prohibitive, and whose scope is often limited. Moreover, from a social point of view, this type of project is often misunderstood by the public and is often "forced through", creating resentment among citizens that is extremely harmful to the image of the developers and to trust between developers and the territories.

Contrary to the common approaches previously developed, the eco-design of a structure requires an integrated and collaborative approach as soon as the objectives of the development are known, that is, at the same time as the definition of the functional and technical needs and before any initiation of the designer's response process.

The approach involves two sequential and hierarchical steps:

1) *management and cessation of pressures*: priority stage of verification of the bio-physical quality of the project areas: uses and activities, mud and polluted water (discharges, over-sedimentation, gray water, etc.), exposure to antifouling[5];

2) *ecological performance adapted to each site*: the objective is to provide functions and habitats for species (prioritizing species under anthropic pressure, or which are "key"[6] to the ecosystems).

This requires a thorough environmental assessment, which should provide a good global and local knowledge of the ecosystem of the site, as soon as the program of the operation is written. This positioning of the development in its "ecosystem" leads the project stakeholders to identify the species, functions and natural habitats of the site and draw inspiration from them to design all or part of the technical, architectural and aesthetic response of the structure to be built. Figures 3.3–3.5 illustrate examples of this demarcation.

As we have already seen, an eco-designed structure integrates ecosystem conservation objectives into its ecological functions in the same way as its technical functions, but in a different order to conventional approaches[7]:

– summarizing the functional needs and constraints of all kinds concerning development within a particular human and ecological environment;

– proposing a preliminary design of the works that meets functional and environmental requirements.

In order to propose a global vision of the eco-design approach, the diagram in Figure 3.6 allows us to understand this integrated design logic.

---

5 Recommended ecological antifouling: ultrasonic hull protection, self-adhesive silicone film ("SilicOne" by Hempel or McGlide), cleaning afloat or ecological antifouling (Nc-coat).

6 Key species: organism whose disappearance would cause too great a change in an ecosystem for other species; it is vital for the functioning of a community (biocenosis).

7 https://www.pianc.org/uploads/files/EnviCom/WwN/WwN-Position-Paper-French.pdf.

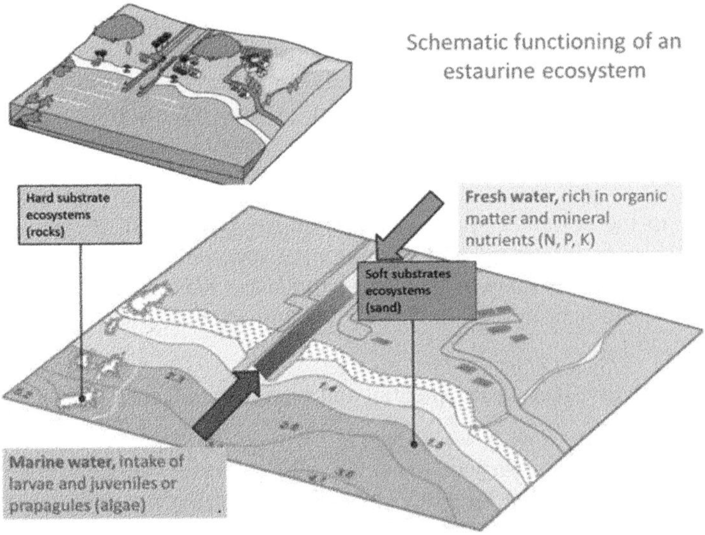

Figure 3.3. *Biophysical functioning: example of a port project in an estuarine area; brackish water (mix of fresh and salt water) is rich in organic matter and nutrients from the watersheds. For a color version of this figure, see www.iste.co.uk/pioch/marine.zip*

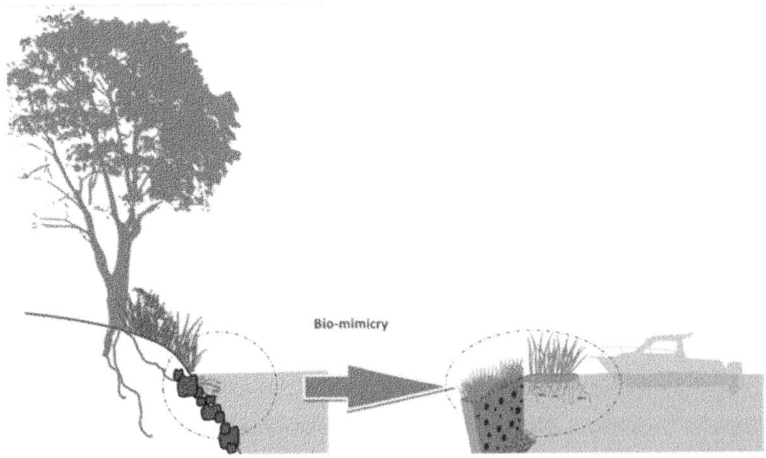

Figure 3.4. *Simplified example of an objective of bio-inspiration from natural habitats: left, submerged natural habitats of a coastal river bank consisting of rocky scree and plant roots; right, proposed eco-design of docks for a river port project, mimicking the size, appearance and shape of natural habitats, for juvenile stage target species (ecological function equivalent with the natural site). For a color version of this figure, see www.iste.co.uk/pioch/marine.zip*

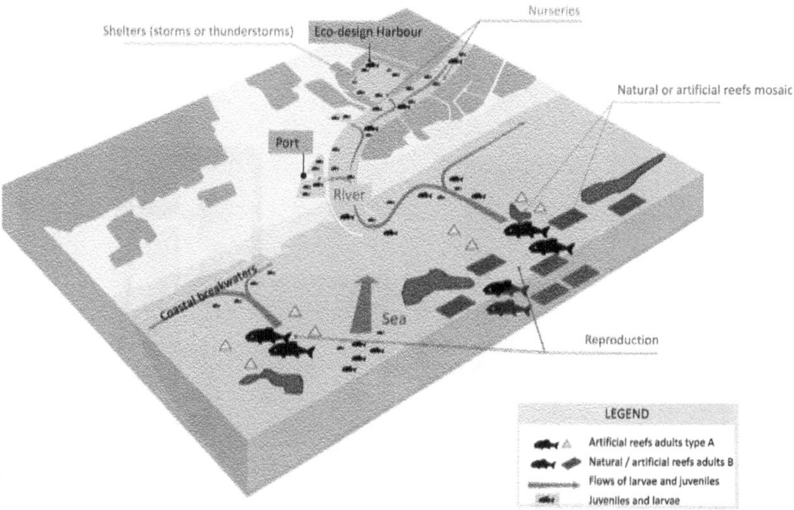

Figure 3.5. *Estuarine ecological functioning: the eco-designed harbor is integrated into a functional ecological landscape. Bottom right and left, breeding areas for adults (spawning grounds, here on artificial and natural reefs in red and yellow); flows of larvae from the high seas, then juveniles when settled on suitable substrates (shelter and food) in coastal shallow water such as breakwaters and in river estuaries. For a color version of this figure, see www.iste.co.uk/pioch/marine.zip*

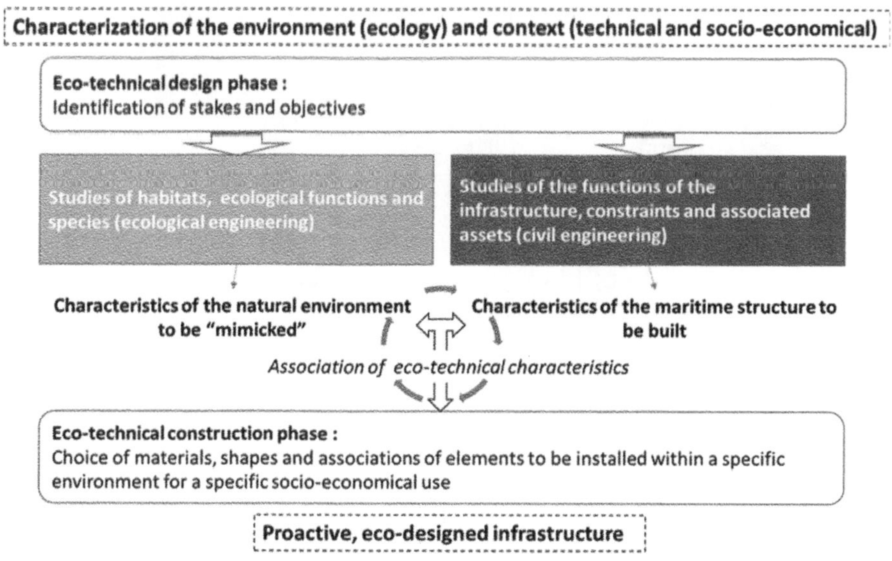

Figure 3.6. *From the concept to the marine eco-design approach (S. Pioch)*

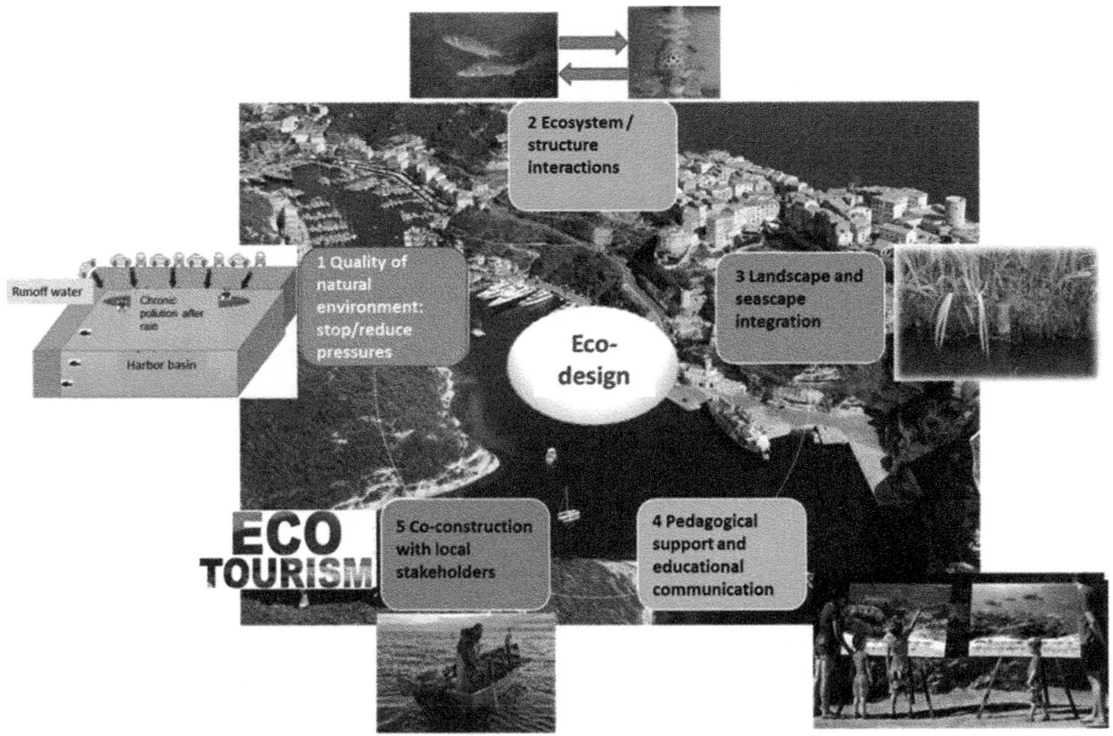

Figure 3.7. *Proposal for the five principles for the implementation of an eco-designed port project: 1) priority to stop and manage pressures on the quality of the natural environment upstream of any project; 2) knowledge of the ecosystem; 3) objective of landscape integration; 4) care given to educational activities or education of users and the public; and 5) a process favoring the co-construction of the project with local stakeholders. For a color version of this figure, see www.iste.co.uk/pioch/marine.zip*

The goal is to propose new approaches and concepts to streamline engineering practices in civil and ecological engineering (Figure 3.7).

At this level of thinking, the question is: *are we capable of doing this for development projects?*

It goes without saying that the eco-design of a technically optimized, economically realistic marine structure with a bio-inspired design presents significant challenges, mobilizing high-level multidisciplinary skills.

However, this is not a pipe dream. It is accessible to those who want it and who have the resources.

3.2.1. *Eco-design actors*

Figure 3.8 presents the development stakeholder involved in an eco-design approach and the relationships that are established during the design and construction process.

One of the main actors is the project owner, because they are the real pilot of the project, giving it impetus and bringing together the actors. The future occupant (often related to the project owner) will have to think on two levels: the functional operation of the structure and the operation of the ecosystem into which the development is integrated. These two aspects will have to be considered when maintenance or rehabilitation works are to be envisaged on the infrastructure. Only the project owner can guarantee the long-term vision from the planning stage to and during operation.

The engineering design office will need to leave its comfort zone, which is that of the standard project management mission (within the meaning of the MOP Act[8] and the Public Order Code[9]). It will need to increase its degree of competence and will be much less able to propose ready-made solutions from a catalog.

State services, for their part, will need to analyze and validate the proposed solutions, in precise national (environmental code, urban planning code and, particularly, the laws relating to the coastline and the reconquest of biodiversity, nature and landscapes) and international regulatory respects, thus favoring realistic environmental approaches.

8 Act no. 85-704 of July 12, 1985, on public contracting and its relationship with private contracting, known as the MOP Act.

9 The *Code de la commande publique* (CCP) resulting from Ordinance No. 2018-1074 of November 26, 2018, and Decree No. 2018-1075 of December 3, 2018 (former *Code des marchés publics*).

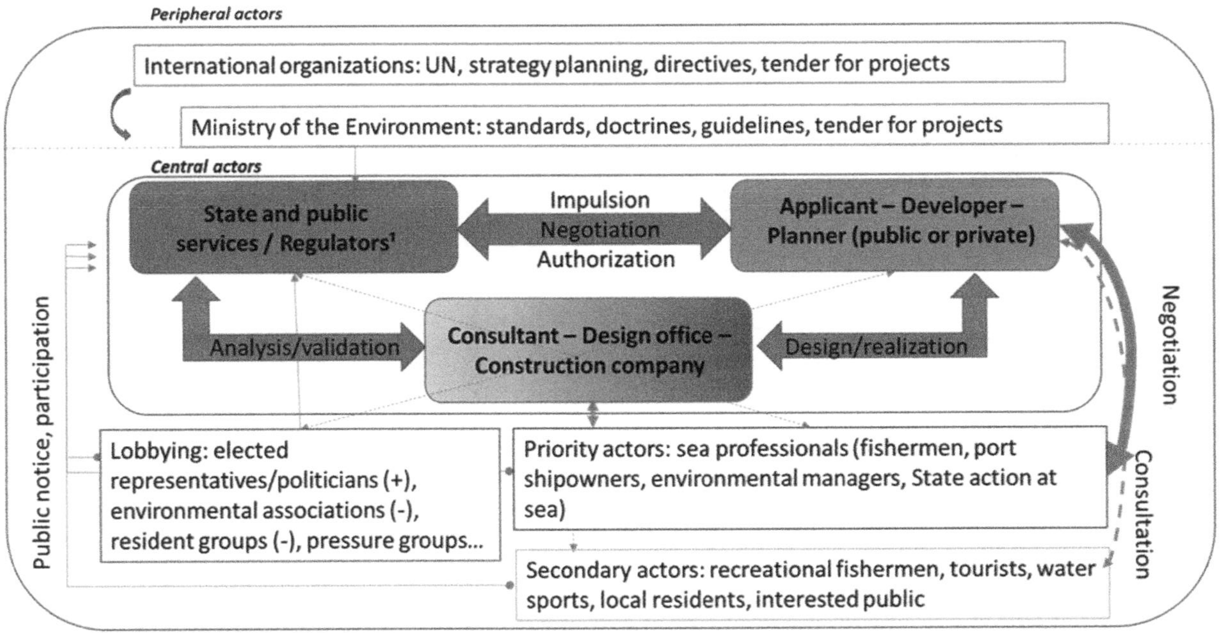

[1] *Environmental authority (Ae), regional or national. With, according to the sites and issues: national council - or scientific regional council - for the protection of nature, national commission for public debate, public establishments (Office français de la biodiversité, Conservatoire du littoral, etc.)*

Figure 3.8. *Stakeholders involved in an eco-design approach and the relationships that are established during the design and construction process at different scales (S. Pioch)*

NOTE.– The logic of selecting the "least expensive" (cheapest) option as a priority criterion for the contracting authority is not compatible with the ambitious objectives of eco-design. The notion of the "best bidder" in the CCP is perfectly adapted. Engineers, in eco-design, can draw inspiration from previous projects by capitalizing on what has been learned, but they cannot be limited to "copying and pasting" standard and common solutions. Lastly, instructing departments must move away from the rigid and restrictive positions of principle that they sometimes adopt. They must offer real support to the project and its actors.

Other actors, such as sea professionals, inhabitants and users, have an increasingly important role in the emergence of eco-designed solutions, relayed by associative organizations or pressure groups, whose lobbying is enhanced by digital methods (social networks, e-debate participation tools) that promote participatory democracy.

Finally, let us not forget the company that will carry out the eco-design work, because it will need to appropriate the eco-design approach and propose execution methods and materials compatible with the initial objectives of the program and with a true creative vision since each environment is unique. At the end of the chain, it is the construction company and the civil engineer who will ensure that the eco-designed solution is feasible and operational from a manufacturing and implementation point of view based on the budget. A reminder here is that the builder executes their professional responsibility with regard to the construction practices in force at the time.

The concept of working with nature, proposed by PIANC, should be put forward. However, to be in the realm of the possible, it is also necessary to have certain convictions about the relationship between humans and nature and about the commitment and the skills that are necessary for this commitment to be enforced.

Today, all the tools for the eco-design of marine structure construction exist, insofar as the actors want it.

3.2.2. *Development and maturation of an eco-designed development project*

Ecological objectives should not overshadow the structural and durability requirements of marine structures, which are always hard-pressed in a marine environment (Souche *et al.* 2016), but put them on the same level.

This horizontal approach (same issues and interests for the socio-technical and environmental aspects) depends on the multidisciplinary skills of the project team:

– engineering sciences and techniques are mobilized to meet the requirements of solidity and sustainability, in an aggressive hydrodynamic and physicochemical context. Ambitious technical and technological objectives are targeted: energy savings, optimization of the use of materials, the circular economy, control of costs and completion times;

– development and urban planning, economic sciences and sociology are mobilized: socio-economic studies, regulation, co-construction of projects and participation of the actors in a common project;

– ecology, ethology and biochemistry are obviously at the heart of the approach: identifying species, habitats and ecological functions, operations and interactions (ecological landscape, behaviors and functions) and the short- and long-term effects of materials on colonization.

In addition, it is necessary to be able to understand and model these different aspects over the long term (operating period), after the development project has been completed, in a rapidly changing context, mainly that of climate change. For this, it is important to produce an environmental cost/benefit forecast.

Environmental monitoring of the structure should be carried out over a period of at least five years, in order to measure the environmental gain in relation to the initial forecasts and objectives. Technical monitoring, on the other hand, is carried out throughout the life of the structure (maintenance and inspection of engineering structures) according to standards and the classical prudential principles of civil engineering, which we will not develop here. In methodological terms, eco-design is systemic, based on an AGILE[10] project development method, which modifies the design process of the structure (Figure 3.9).

In such an approach, it is necessary to consult the different actors and inhabitants/users of the project territory (Figure 3.8) from the beginning of the operation. In the interests of the efficiency, functionality and sustainability of the design, it is imperative to integrate these actors in the different phases of the project, especially during the design phase. This can be achieved through consultation or follow-up meetings, digital communication facilitating debate and direct questioning

10 The AGILE method consists of setting objectives and providing solutions. Then, iterations are indexed according to the situation at time "t" until the completion of the final structure. It is an empirical, iterative, transparent approach that involves all the skills present in order to meet all the requirements with optimal risk control.

of the project owner (e-debate). Active listening approaches (reformulation of questions) help to bring out the "brake" and "driver" arguments of each (fishermen's associations, researchers, companies, communities, etc.). The search for compromise should be the rule.

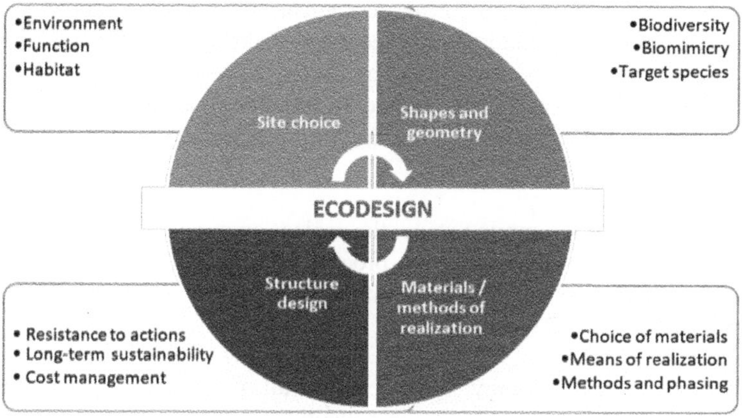

Figure 3.9. *The eco-design approach, a systemic, AGILE approach. For a color version of this figure, see www.iste.co.uk/pioch/marine.zip*

NOTE.– The compromises adopted must always keep two main objectives in mind: a pragmatic, realistic, technical approach, adapted to the marine environment; and effective ecological solutions, adapted to the study site.

Generally, for a public investment, the first phase of programming is the definition of the requirement. The project owner must make sure that they will control the PMD and the *land* necessary for the construction of the structure and that they can set up a *finance plan* adapted to the operation (work costs, study costs, land acquisition, etc.).

Eco-designing a development requires having the right data and hypotheses from the preliminary studies and in particular defining clear environmental objectives. The preliminary studies must therefore specify the requirement from a technical and environmental point of view, and it is on the basis of an initial ecological assessment that the functioning of the ecosystem is understood. It is therefore essential to have at least one ecologist and one civil engineer in the team of consulting engineers assisting the project owner. Other development specialists (urban planners, landscape architects, economists, etc.) may also be called upon. It is from this stage that the public is consulted and, ideally, involved in the co-construction of the project (AGILE method or similar).

The project owner then formalizes all the programming elements of their operation in a framework document called the "program", which contains the specifications for the development and construction operations. This document sets out all the technical and functional components of the project, as well as the objectives to be achieved, including from an environmental point of view. It integrates the administrative, technical, ecological and financial feasibility, as well as the process chosen for the correct execution of the project. On no account can eco-design, which consists of adding a strong ecological functionality to the structure, be interpreted as reducing its overall functional performance. The program will be the reference throughout the construction process and will be a key element at the start of the operation.

To complete the programming stage, the project owner chooses a methodological process for carrying out the construction process: organization of the actors, sequence of tasks, identification of the regulatory procedures for technical studies, environmental studies and work, etc.

The eco-design methodology, which enables all of the actors to be coordinated around the project, is highly effective in consolidating the construction process and minimizing the risks of administrative appeals throughout the project. Figure 3.10 summarizes the approach of the project owner who integrates eco-design during the programming phase.

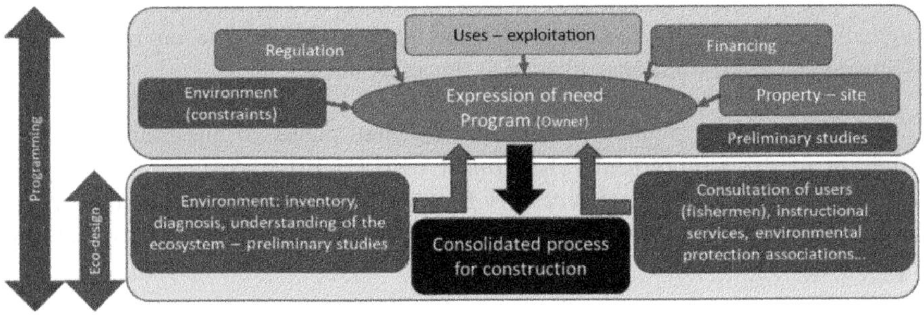

Figure 3.10. *Project owner programming stage, supplementary to the eco-design. For a color version of this figure, see www.iste.co.uk/pioch/marine.zip*

This raises the question of the interest of the project owner in voluntarily choosing an eco-design of the structure. Is this not an additional complication for the project?

Indeed, there are always difficulties in adopting an approach that integrates the environment. Generally for a PMD, projects are carried out by a public body and it has legal obligations to open its markets to competition and to guarantee equal treatment of candidates. These rules, which are pillars of the public procurement code, often lead to segmenting and compartmentalizing the different actors, which is detrimental to the systemic approach advocated here. Moreover, when innovation is at the heart of the project, the usual procedures for awarding public contracts are often restrictive and unsuitable. It is therefore necessary to adopt alternative procedures that are still not widely used by the procurement departments of local authorities, for instance.

At first glance, for the project owner, an eco-design approach adds complexity to the programming of an operation. Two important points can change their view and encourage them to favor an eco-design approach (Agropolis International 2014).

As demonstrated by the testimony of the Mediterranean Prefecture (section 4.7.2), firstly, there is a regulatory reinforcement in favor of the protection of coastal biodiversity and coastal ecosystems in development works, as well as for the restoration of natural functions. This is in response to the unprecedented crisis we are experiencing.

Eco-design is becoming increasingly important under the effects of various European or national regulations. Its application is becoming increasingly widespread and is less and less voluntary. A recap on relevant policies: there is the Habitats, Fauna and Flora Directive (DHFF), the Marine Strategy Framework Directive (MSFD), the Marine Environment Action Plan (PAMM), the law on the reconquest of biodiversity, nature and landscapes, etc. Eco-design thus becomes an argument that highlights the virtuous approach of the project owner to the instructing services, but, also and above all, to the public, which can rally around the project.

This approach is also encouraged by certain project financing organizations. Indeed, this eco-design approach is beginning to develop at the national level as a result of recommendations and public aid. For example, the Loire–Brittany Water Agency covered 50% of the additional costs relating to the implementation of eco-designed works in its 2010–2015 five-year plan. Similarly, the charter of the Gulf of Lion promotes the approach by supporting this integrated vision of marine coastal development. For mooring projects, the Corsica environmental office mainly finances projects that include an eco-design component. Any eco-designed project can therefore be eligible for funding.

The second real interest for project owners is linked to the integrated vision of the project design, by opening up the discussions to a wider audience in order to respond to the different actors of the marine environment, including professionals, institutions, citizens and associations. The involvement of these different actors, from the very first phases of the design, is an excellent way to avoid discovering environmental problems, or social opposition to projects, at a later stage.

This can reduce the risks that weigh on the project and enables the project owner to better control the project schedule. Eco-design is a strategic approach whose very purpose is the success of a project integrated into its environment, as highlighted by the testimony of Cyrille Taioni, director of port operations at the regional port of Sète (Box 3.3).

Finally, let us recall a direct benefit to structures that has been identified by some recent articles cited above: the living organisms that settle on structures could offer bioprotection to the structures and thus improve their sustainability (Coombes *et al.* 2015).

While eco-design offers an obvious environmental benefit to a structure, perhaps tomorrow, it will be possible to quantify a technical benefit, by improving the durability of structures at sea. This would be a "win-win" scenario.

"In 2014, the Languedoc-Roussillon region, under the presidency of Christian Bourquin and the vice-presidency of Didier Cordorniou, created the 'sea parliament'. This innovative consultation body brings together 200 members from the marine community, users who previously had no reason to meet: fishermen, biologists, ship owners, nautical professionals, port managers, civil engineers, historians, scientists, etc.

On the occasion of a workshop on eco-design, organized by the sea parliament, I expressed concern about this approach that aims to provide habitats within a civil engineering structure, which I know that we must by definition maintain. As a good manager of a complex port heritage that is the port of Sète, whose first structures, of which 80% are under water, date back to 1666, I reason as a technician that the less life there is on a structure, the easier it will be to intervene to repair it. However, the concept is gaining ground: why build a structure that will be inert, when we can improve its ecological role? It will be necessary to make the link with my terrestrial experiences to understand the interest. On the roads, geomorphological slopes have enabled a colonization of vegetation thanks to natural steps, which retain Earth and water; or again, architectural matrices, which, integrated into the design, have allowed for a lasting and unobtrusive effect on the eye. In both cases, by modifying conventional methods, a sustainable, resilient objective is achieved.

The meeting with Jean-Claude Souche and Sylvain Pioch did the rest to convince me definitively. They are going to dive and bring me back pictures showing the life that has settled along a dock over a period of well over a century that we have to rehabilitate because the concrete is crumbling and cavities are being created here and there. The fauna and flora have taken advantage of this decay and have adapted to these shelters of opportunity.

How do I repair this dock, now that I know there are many species on it? How do I provide for the right resilience of nature on this dock which is full of life, in the face of the need to repair it? By artificially recreating habitats similar to those that time has shaped? How do we make this habitat sustainable?

The proposed solution is to carry out the repairs using an elastomeric formwork skin and schedule work to reproduce the habitats present today. From a technical point of view, it will be necessary to ensure that the concrete steel cover always remains higher than 5 cm. How do I make this project financially viable? Through the principle of integrated ecology, the additional costs become marginal at the investment stage and have no impact on maintenance. This project must be completed in 2021–2022; the key words that prevail in this reflection are: resilience, sustainability and design integrated to the environment.

As a project owner, I have a request: I would like to see the regulations evolve so that if a protected species were to be found on a structure, repairing it with a dock which is eco-designed could constitute a measure recognized as compensatory."

Box 3.3. *What are the reasons for a client to impose eco-designed works? (source: Cyrille Taioni[11])*

3.3. The methodological approach to eco-design: responding to the expressed need

As seen previously, the eco-design of a development requires a preliminary environmental study. This section specifies a few points that should not be overlooked during this first phase of the environmental engineer's work with the civil engineer.

As a reminder, the first step in eco-design is, of course, to ensure that the pressures on the natural environment are managed or controlled.

It is very important to emphasize here that this approach is to be carried out in parallel with the ARO[12] approach in the impact study, a study that comes later (in the pre-project, as the preliminary design, for example, in the French context).

The second stage of eco-design is the description and analysis of the natural biophysical environment.

11 Head of the port operations department at the Occitanie region, Sète.

12 Avoid, Reduce, Offset (see Chapter 1).

The natural marine environment is characterized by:

1) physical parameters, such as: currentology, sedimentology, thermocline, salinity, turbidity or topo-bathymetry;

2) chemical parameters, which must also be considered: they can particularly testify to a pre-existing pollution or to the eutrophication state of the environment;

3) ecological parameters, determination of which is one of the most important phases. It consists, among other things, of determining the species and colonizations present, the habitats and the ecological functions (list of species, identification of relevant or key species, biocenoses map, quality and map of habitats, zonation of functions, past trajectories and projections in a probable context of climate change).

These data are of course identified and analyzed in any project, but usually in the initial design phase (preliminary studies and/or pre-project) as soon as the choices of site and proposed techniques are put in place. We can remind ourselves here that, for an eco-design approach, they must be available from the very first phases of the project design, well before the so-called "environmental" studies relating to the environmental assessment (environmental impact study). These elements are identified in addition to the "classic" data relating to socio-economic, regulatory, local and institutional aspects.

The consideration of biophysical parameters at the beginning of the project, with a precise ecosystemic vision, is undoubtedly the key element to the success of an eco-designed approach.

3.3.1. *Eco-design, an adapted technical response for engineers and project managers*

We have seen that the approach of the project owner is closely linked to the decision of whether or not to eco-design a development. What about the engineering firms that act as project managers?

In France, private project management for public clients is governed by the MOP Act, which specifies the content of project management assignments. Elsewhere, in other countries, other relatively similar texts propose the same principles.

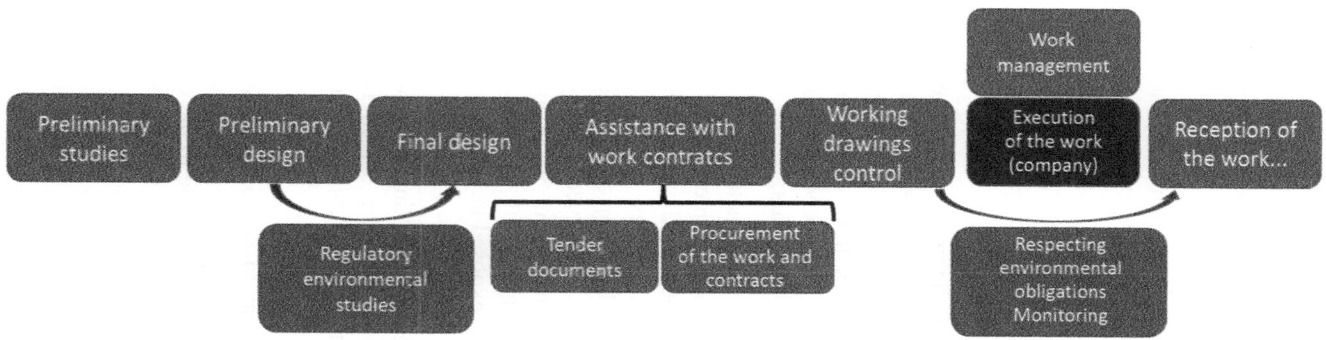

Figure 3.11. *Usual project management assignments for a marine engineering office (project engineering according to the MOP Act). For a color version of this figure, see www.iste.co.uk/pioch/marine.zip*

Figure 3.11 shows the usual project management tasks carried out by engineers in design offices (civil engineering, environment, socio-economic, etc.). The regulatory environmental studies, which are necessary to obtain the authorization to carry out the work, are generally carried out during the pre-project phase.

Aside from its conceptual originality and the lack of dissemination of good practices, taking the natural environment into account from the start of the preliminary studies changes the designer's point of view on several levels:

– taking into account environmental objectives requires a local vision of the structure as a stakeholder in the ecosystem. For each project, it is therefore necessary to start with a blank canvas and the "copy and paste" of solutions previously implemented remains very limited;

– there is a twofold aspect to the time scale of the structure's operation: from the point of view of civil engineering, and of long to exceptional periods of use (30, 50 or 100 years); from the environmental point of view, from the almost immediate colonization stages by biofilms (less than 1 hour, made up of cyanobacteria and then diatoms), to the long equilibrium between species (competition, speciation, selection) over tens, even hundreds of years;

– the geometry ("architecture") of structures will depend, on the one hand, on the constraints linked to the operation, the mechanical requirements of resistance and load bearing, and, on the other hand, on the ecological functions, the types of habitat to be mimicked, and the target species, for a better integration into the local ecosystem. The compilation of these objectives imposes the permanent search for compromise, since the environmental objective may collide with the objectives of sustainability or three-dimensional form: for example, the creation of holes or vertical alveolar forms for the purposes of sheltering species would be incompatible with the stability of a structure exposed to a strong agitation requiring massive and low forms;

– the choice of materials is not only linked to the sustainability requirements of structures. It is also associated with the search for better colonization capacity through the life and realization of biomimetic structures (surface aspect, color, orientation, pH, composition, etc.).

These objectives may appear to be contradictory because a concrete that has been considered "sustainable" until now is compact, closed and smooth on the surface, whereas a concrete that is favorable to the colonization of sessile species is porous and rough. However, the designer can positively respond to eco-design objectives and find the right compromise under two conditions: (i) that this is taken into account from the initial stages of design (in fact, from the start of the program, etc.); (ii) that the work is carried out as a team with the common goal of technical and environmental optimization of a development integrated into its environment (Figure 3.12).

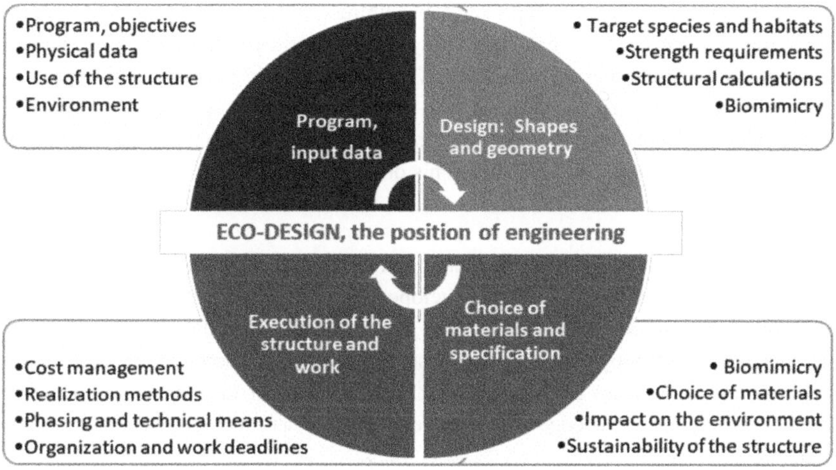

Figure 3.12. *The eco-design approach for engineering, a systemic and ambitious approach to the search for the best compromise. For a color version of this figure, see www.iste.co.uk/pioch/marine.zip*

It goes without saying that the eco-design of a beautiful, technically optimized maritime structure integrating a biomimetic design represents a significant technical challenge and must mobilize the high-level skills of a multidisciplinary team with a real ability to listen to each other. The purely financial or organizational vision of the "project manager" is not enough to meet this challenge.

What does Romaric Vicente, a specialist engineering manager, think? His thoughts are given in Box 3.4.

"The role of engineering companies has always been to solve complex and concrete technical problems in the service of society's requirements. As a result of this, they have developed know-how that enables them to meet criteria of reliability, technical performance, economic competitiveness, risk management and sustainability. The engineers who make up their main workforce, as well as the schools that train them, have always focused on developing the skills associated with these challenges.

The environment, as such, was for a long time considered as an external factor influencing the choice of technical guidelines adopted in the design of projects. The phenomenon of climate change, as well as the other consequences of human activity on the environment, now places the criterion of societal and environmental responsibility of proposed structures at the center of everyone's concerns. The first standards relating to eco-design appeared at the beginning of this century and expectations on these themes are growing.

Although environmental criteria are becoming increasingly important in project development, they are still too weak compared with purely financial criteria. Nevertheless, the trend is there and engineering companies, as well as all actors in the construction process, are progressively integrating these concepts into their thinking. These criteria require finding innovative technical solutions that take the environment into account and create value for society, in contrast to an approach that has historically often been considered restrictive and limiting.

Among the different approaches originating around eco-design, biomimicry appears very interesting. The observation of solutions developed and selected by nature allows engineers to be inspired by *sustainable* solutions proven by the evolution within the *biosphere*. However, a simple morphological copy can lead to ill-suited solutions. Therefore, it is necessary to approach the problems from a systemic point of view in order to understand all the factors involved. This implies knowledge of components that are as much physical as biological or even chemical.

In the end, these approaches place the engineer back at the center of the debate on the need to invent new ways of designing our structures. The engineer thus acts at the heart of the environmental transition in the name of: (i) the preservation of biodiversity; (ii) respect for our finite planetary resources; (iii) more broadly, the sustainability and quality of life of humanity.

In this respect, engineering, and more specifically the engineer, has a key role to play in enabling enlightened and concerted societal decisions in a complex environment, often torn by contradictory factors. Through eco-design, a real paradigm shift is taking place in the way we think about our structures for the benefit of the common good."

Box 3.4. *The changing nature of environmental consideration. How should engineering firms and the design engineer evolve in order to eco-design? (source: Romaric Vicente)*[13]

Eco-design thus places an innovative, technically advanced and entrepreneurial design engineer back at the center of the game. We return to the fundamentals of illustrious engineers such as Gustave Eiffel or Eugène Freyssinet[14], who at the end of the 19th century had new technological tools at their disposal and were faced with the new technical and societal challenges of an industrial revolution: new materials (cast iron, steel, cement), large foundries, worldwide deployment of railways, development of cities, etc.

13 Deputy Director of the company BRL Ingénierie.

14 https://www.linternaute.fr/science/biographies/.

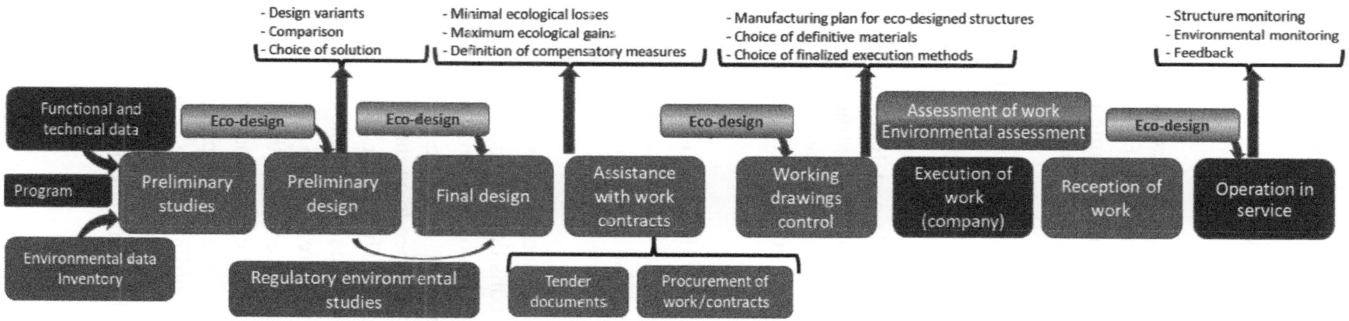

Figure 3.13. *Adaptation of project management assignments for the consideration of eco-design through the engineering of marine developments. For a color version of this figure, see www.iste.co.uk/pioch/marine.zip*

Technical skills are becoming the driving force behind a reasoned and pragmatic innovation that makes it possible to respect the usual criteria (economic and environmental, risk management, compliance with standards, etc.) and then to advance beyond them to develop brand new "tailor-made" technological solutions: *the eco-design of marine structures*.

All this might seem very feasible, but it is much more complicated when the imposed deadlines and the political stakes push engineering firms to confuse speed with haste in order to respect their own economic result criteria. To repeat, the aim is not to wipe the slate clean for the past, but to modernize the engineering approach to marine development. Eco-design is a real opportunity for engineering firms to reposition themselves as major players at the heart of the construction process and at the center of the game.

Project management must also evolve. It will no longer be possible to continue to entrust engineering studies to the lowest bidders, for whom price is the main focus, while at the same time hoping to build structures with ambitious, innovative and "eco-"efficient objectives. The construction of a technically optimized marine structure integrating a bio-inspired eco-design, including an underwater aesthetic vision, is an illustration of this.

Figure 3.13 puts into perspective the methodology that an engineer will need to follow to achieve eco-design of a marine structure. The interdependence of the key actors in the construction process, that is, the client, the contractor and the company, is emphasized.

It is also necessary to take into account one final specificity of the eco-designed marine structure. At sea, if we consider the "colonization" aspect, things are quite different: each square millimeter is covered with a biofilm (only a few hours after immersion, made up of cyanobacteria and then diatoms), then with an assembly of more complex flora and fauna that depends on the depth, the seasonality and the needs of the living organisms considered. The height of a dock will not be treated in the same way close to the surface or at the base of the dock, since the target species will not be the same. The engineer therefore needs to zone the infrastructure in terms of the forces to be withstood, the aggressiveness of the environment and the capacity for colonization. This is an additional 3D effect brought about by taking into account the reciprocal effects of ecosystem and development. Finally, the seasonality and temporality of subsequent years contribute to the development of a rich marine life and thus to induced uses of the structure. These factors create the 4D effect mentioned at the beginning of Chapter 3, an effect that must be taken into account during the design phase.

Ultimately, eco-design provides the engineer and designer of "marine" structures with an ethical responsibility. Beyond the response to societal and functional considerations, they can act for biodiversity by simply using a "fertile" aquatic environment associated with their technical skills, and they must do this at a fixed cost to avoid falling at the first hurdle with regard to the project budget.

This is the very essence of sustainable development, which the designer can also associate with the new and important concepts of the circular economy and optimization of the use of resources. If it were not already enough for their role to be useful, the role of the designer now becomes "noble", because they act daily for their environment, for the sea and for future generations.

Thus, the ecological added value sought would improve the technical performance of structures. The space left to nature would improve the human endeavor, and the human endeavor would be closely linked to that of nature – a marriage of convenience.

3.3.2. *Eco-design as a lever for modernizing the execution of works*

Due to the resources employed, execution of the work can have negative and heavy impacts on the environment. It is therefore imperative, very early in this phase, to identify the impacts of the work and to find measures to minimize them, or, failing that, to reduce them and, as a last resort, to compensate for them. For example, on the site of the new coastal road in Reunion Island, the phasing of the mining and excavation was adjusted so that the work could be carried out outside the whale breeding period.

In the case of marine works, the installation of new equipment into the environment causes abrupt changes to the habitats and sometimes generates the suspension of sedimentation. To repeat, development always involves sources of disturbance and sometimes pollution for ecosystems; but even if this cannot be managed without destroying nature, the effects can be considerably reduced, depending on the technical choices that are made.

There is a real awareness in construction engineers and they too are realizing their role in this new way of seeing things. Frédéric Martarèche, a graduate of the Ecole Centrale Paris, former civil engineer and then manager of the company Bec Frères SA, testifies to this in Box 3.5.

"After graduating with an Engineering degree from the École Centrale Paris, I have spent my entire career with the publicworks company BEC Frères SA, successively as a methods engineer, then civil engineer, then as a freeway site manager, a subsidiary manager in Southeast Asia and a project manager for the Monaco dam, and then, as a member of the management committee, as an international manager and technical development manager. I have therefore contributed to the construction of large infrastructures: freeways, dams, canals, large concrete structures, quarrying, etc. I have always been aware that the very activity of public works has a strong impact on the environment and that it was therefore essential to integrate these aspects as much as possible in my management roles.

In fact, in the 1970s and 1980s, the landscape integration of infrastructures and the reduction of their impact on biodiversity were the concerns of a few enlightened pioneers! Much progress has been made in this area. Today, no developer would imagine proposing a land-based structure that would not be attentive to environmental considerations. Besides, such an indelicate developer would obviously not obtain the required authorizations to build.

As the company's technical development manager, I was involved in the creation of the experimental eco-designed reefs at Cap d'Agde in 2009, in partnership with the École des Mines d'Alès and the engineering company Egis Eau. It was then that I met Sylvain Pioch and Jean-Claude Souche. This first experience in a marine environment enabled me to see first hand the interest in eco-design for marine structures and the enormous potential for improvement that is within reach.

On the other hand, I was able to observe that, in the marine environment, the consideration by developers of the impact on the environment and biodiversity was 'late' and not yet fully established as common practice, unlike the practices of infrastructure on land! It must be said that, for a long time, we believed that the ocean was infinite and that, as such, human developments on the coast were virtually without consequence. Fortunately, we have now collectively begun to understand that marine biodiversity is highly impacted by human activity.

Aware of the criticisms of environmental damage from public works, I have always been one of those convinced that it should be possible to reconcile human activity and biodiversity, provided that this is integrated into the design of structures and during the execution of the works.

Thus, in 2013, after leaving BEC Frères, the opportunity arose for me to participate in the creation of a start-up within the EGIS Group (Seaboost), to which I could bring my entrepreneurial experience of methods, cost estimation and construction organization. As a 'wise old head' within a young team of engineers and divers, I could therefore use the experience acquired during a career in France and abroad in the service of my deep convictions. I could participate in a new approach reconciling public works with nature. Convinced that there is a real and urgent need to systematically integrate eco-design into any human intervention in the marine domain, I first had to make sure that this 'very fashionable' subject was not just a passing fad, but that it met a need and a market, while avoiding the trap of green washing, as the stakes were too real to imagine cheating and deceiving others.

The strong idea is that a development, beyond being a source of degradation of the environment at the time of the work, may also be considered as a habitat opportunity, participating in a new ecological balance with the environment in which it is inserted, especially if biomimetic objectives are sought. Knowing that economic validation remains sometimes difficult to demonstrate, it is necessary to develop the concept that any marine infrastructure must contribute to the preservation of biodiversity;. especially since, very often, the additional cost is very low if the measures are incorporated as soon as possible into construction budgets, as is the case in land infrastructures.

On the other hand, it is necessary to help clients to justify that the expenditure is judicious, and henceforth unavoidable. I am convinced that the marine construction company has a place in this process, where its competence, resources and the methods it employs are major elements in the economic justification of the structure and in the impact of the site on the environment."

Box 3.5. *Frédéric Martarèche's personal vision*[15]

In marine works, the resources used for the pile-driving of foundations or for the dredging of a port, or for the lifting of civil engineering parts, are often heavy resources, adapted to large works of immense infrastructures to be built and which are proportional with the expected technical functions, particularly the sizes of ships to be accommodated.

The famous landscaping adage, "big project, big impact", is a fact, but not an inevitability. In fact, it is possible to promote clever techniques and methods that minimize on-site interventions. For example, the prefabrication of concrete elements can reduce the time required for on-site intervention. The choice of a suction dredging method can reduce the quantity of suspended fines, compared to a dipper-dredge barge (mechanical shovel on a pontoon).

In some cases, it is necessary to try to give the ecosystems in place the opportunity to adapt and move to other habitats that may be available to them. For example, soft start or ramp-up techniques allow, through a gradual and progressive start of work operations, vagile species that are able to move to be kept away. These

15 Former construction manager for the company BEC Frères, reconverted to the eco-design of marine works within the company Seaboost.

considerations inevitably lead to a greater complexity in the methods and stages of construction, requiring greater technical and organizational skills.

On the other hand, the origin of the materials used, the place and the manufacturing techniques of the constituent elements, as well as their transportation, should be considered in terms of energy expenditure (carbon footprint) and in terms of local and global scales, in particular via the life cycle assessment (LCA) (Langlois *et al.* 2020).

Materials that are close to the site should be favored as well as bio-based and non-polluting materials. For example, for metallic structures, the installation of cathodic protection by sacrificial anode (an alloy using mainly zinc, aluminum and magnesium as well as indium and cadmium) results in the release of heavy metals into the environment, during the life span of the anticorrosion protection. Can we not promote alternative methods which could see living organisms providing a physical barrier against corrosion and protecting the structure (Figure 3.14), as recent research work tends to suggest?

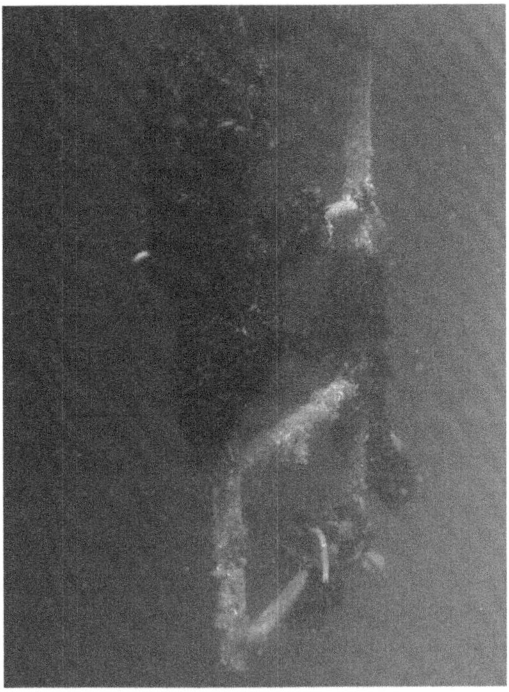

Figure 3.14. *Port of Sète, dock H, presence of mussels on a metal ladder (photo: J.-C. Souche)*

Similarly, in parallel and more classically, the recycling of materials should be considered in order to minimize the consumption of natural resources, in accordance with the principles of the circular economy.

For certain structures, the deconstruction of the infrastructure and the restoration of the site must be integrated into the design of infrastructures when they reach the end of their service life, requiring cross-planning with the available documents: the territorial coherence scheme (marine section), the sea development plan, the master plan for water development and management, etc.

These legitimate concerns must be taken into account as far as possible, along with the necessary financial optimization of the project and the technological feasibility of the measures proposed, which cannot contradict the elementary rules of common sense.

3.3.3. *Monitoring, control, validation and satisfaction for the user and nature*

As we previously mentioned, during the different phases of the development project, eco-design, as an integrated systemic vision, requires the consideration of multiple issues in collaboration with the actors and wider public of the marine territories involved. Transparent communication of the functioning of a structure, in a natural environment, becomes, in effect, essential in our eyes, at the risk of blocking out a society concerned about the future of its unique planet. This is true during the study phase of the project (design), during construction (works) and naturally during the whole life cycle of the work (operation). Once the work has been completed, it is of course normal that the actors concerned by the operation of the work take on important roles.

The operation of a structure may have negative consequences on the environment or on the structure itself, or may generate additional maintenance costs. It must also be ensured that the provisions made for biodiversity do not lead to negative "side effects" – such as the example of hollow blocks positioned in dike shells and subject to significant sediment deposition, and therefore requiring regular maintenance – or the solution will be declared invalid (Waltham and Sheaves 2018).

The issues and risks are therefore identified at the design stage, but it is necessary to consider how they will change over the life of the infrastructure and how best to respond to them. For example, it is possible to create a management structure, composed of different representatives interested in the operation of the infrastructure and concerned about its effects on the natural environment. It could then take care of the implementation of rules of use or recommendations, and then monitor their respect according to the expectations of each, to ensure the sustainability of the structure and the activities related to it, as well as the environmental consequences of the infrastructure throughout its operation.

On the other hand, a charter of use can allow for better cohesion among the actors and the public. It is also possible to think of a monitoring plan self-managed by the users and involving interested independent external parties (fishermen, environmental associations, experts, etc.). The main thing is to provide a usage framework and to clearly define the rights and duties of each party, with regard to technical-economic as well as environmental objectives. For example, Ruhl and Craig (2009) have proposed new governance contracts for coasts and estuaries. One of the models proposed for integrating a sustainable development policy is *integrated adaptive ecosystem management*, where the ecosystem is managed as a whole, including human activities. The goal is to maintain a healthy ecosystem, producing the expected ecosystem services for human needs and with a sufficient level of resilience. These new modes of governance are envisaged because the usual management systems (integrated coastal management or estuary management) do not take into account the complexity of marine systems, where there is a strong coupling between natural resources, human activities and needs and the sustainability of ecosystems.

The purpose of all of this is not to place new environmental brakes and constraints on the operation of the structure, but rather to operate in an informed and known manner, so that the functions of operation, maintenance and eventual rehabilitations of structures can be facilitated, or even improved due to the permanent consideration of ecosystems and their interactions with infrastructures; for it is absolutely necessary to incorporate that a marine infrastructure must be maintained and, as such, be subject to maintenance work. These interventions will have to be considered and carried out according to the same eco-design methodology as the structure itself, with actors who have the same constructive and open mindset.

Today, some public organizations have adapted and are leading the way. They have perfectly understood the stakes related to the environment, and are leading a proactive, voluntarist and courageous policy. This demonstrates the relevance and feasibility of the concept. They are assuming a new responsibility for any 21st century planner, the destiny of humans being directly and intimately linked to nature, of which we are a part and of which we need to "be".

3.4. Infrastructure as a new support for marine life

In this section, we propose to apply the conceptual and methodological contributions of both environmental assessment (environmental impact assessment) and ecological engineering. In order to remain as close as possible to the aspects that we consider innovative in the eco-design approach, we will not define in detail here the technical steps needed for a good knowledge of an ecosystem and the establishment of an environmental initial state, which are available in an abundance of modern publications[16].

On the other hand, the quality of the methods and tools used to collect these data is absolutely fundamental, based on essential aquatic observations carried out by specialists in marine ecosystems using snorkeling or scuba diving (if possible with air rebreathers rather than diving tanks). Without this, no significant long-term ecological improvement of structures will occur. We will first look at the abiotic, biotic and landscape factors that allow us to define and understand the functioning of an ecosystem, then at their use in an eco-design objective according to several scales ranging from the infrastructure in its environment to the colonization of the chosen material.

3.4.1. *Biophysical data of the environment*

In order to integrate an infrastructure into the natural environment, it is necessary to study the biophysical, biotic and abiotic functioning of the environment. This is what defines the ecosystem, which is the whole formed by a community of living beings in interrelation (biocenosis) with its environment (biotope).

3.4.1.1. *Study of abiotic factors*

While the functional analysis of colonizations provides information, notably on the trophic interactions between organisms (who eats whom?), the physical environment (abiotic) concept provides elements of understanding on the interactions that these same organisms have with their habitat (who lives where?). This physical dimension of the ecosystem constitutes the biotope of the organisms monitored and largely conditions the distribution of the biocenoses previously studied. However, the study of this physical environment requires the skills of a multidisciplinary team, as the number of environmental components constituting the biotope is high.

As with the study of biotic factors, this requires the implementation of specific sampling and analysis techniques that can be complex. These elements were discussed in Chapter 2; as a reminder, they are factors relating to (i) climatic and

16 We recommend to readers the *Guide pour les études d'impact environnemental (EIE) en milieux coralliens de France outre-mer* by Mathieu Pinault, Sylvain Pioch and Nicolas Pascal from the *Documentation Ifrecor* series, published in 2017. Available free online at: http://www.ifrecor-doc.fr/items/show/1742 (accessed June 24, 2020).

telluric influence; (ii) tidal range and hydrodynamics; (iii) bathymetry and geology; and (iv) the physicochemistry of water and sediments.

3.4.1.2. *Study of biotic factors*

Knowledge of the biocenoses, the "grouping of living beings whose composition, the number of species and that of individuals reflect certain average conditions of the environment […] linked by a reciprocal dependence", is essential. The assembly formed by an environment (biotope) and all the species (biocenosis) that live, feed and reproduce there constitutes the ecosystem (biome).

Each species, whether animal or plant, is characterized by its way of life (occupation of space, rhythm of activity, etc.), its requirements (food, physiological) and its adaptive potential (resilience). The sum of these characteristics defines its ecological niche. It is therefore not simply a matter of location and use of space; the ecological niche represents the role of each species within the ecosystem. This concept allows us to better understand the relationships (competition, cohabitation, association, etc.) between species in the ecosystem and to interpret certain biological evolutionary mechanisms: the occupation of new ecological niches can effectively promote the appearance of new species (extension of the natural distribution area of indigenous species or colonization by exotic, possibly invasive species). These niches may also be interconnected, either by direct links between juxtaposed habitats (continuous ecosystem) or by the relay of ecological corridors (fragmented ecosystem).

In summary, the study of biocenoses is expressed around three complementary components (Figure 3.15):

1) the compositional description of colonizations, which includes the number, composition and heritage interest of the species that make up the biocenoses;

2) the structural characterization of communities, which focuses on defining and understanding how the elements of the colonization are organized in relation to each other;

3) the functional analysis of biological processes, which through two essential points, the quantification of individuals and their interdependence (feeding, predation, association), innovates in a decisive way with regard to the so-called "fixist" (without dynamics, nor evolution) descriptions hitherto given to biological communities.

These three components of biocenoses are discussed in sections 3.4.1.2. 1–3.4.1.2.3.

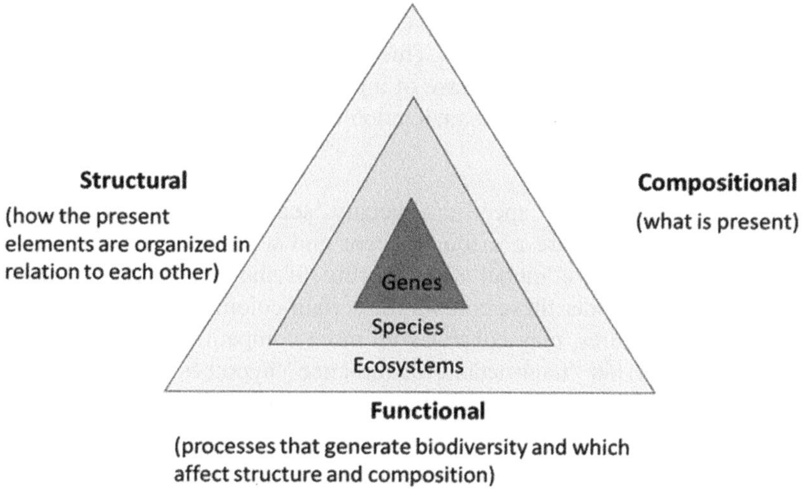

Figure 3.15. *Compositional, structural and functional approach to a biocenosis (Pinault* et al. *2017)*

3.4.1.2.1. Composition of colonizations

Among the criteria for evaluating the issues relating to the natural environment and biological equilibrium, the compositional description of the colonizations is probably the most immediate and the most frequently used in environmental impact studies. Biocenoses are formed by a number of species that can be very high, exceeding 1,000 or even 10,000 in complex environments such as coral reefs. It is the coexistence of all these species in the same small area which characterizes the specific diversity. A biocenosis is an extremely vague entity and, consequently, its specific diversity is very difficult to define, which rather obliges us to consider it at the level of certain taxonomic groups (fish, corals, algae) or certain communities (plankton, sedimentary macrofauna, etc.) that make it up and to reduce it to the sampling effort that was necessary for its identification (time, surface, volume). A second difficulty lies in the interpretation of specific diversity values. While it may seem logical that a stable and preserved environment should contain a greater number of species than the same environment once it has been disturbed and degraded, the reality is more subtle to decipher.

It is recognized that a structured ecosystem with a good conservation status contains a greater number of species than an immature or degraded ecosystem, which is subject to frequent and high intensity impacts. On the other hand, no linear relationship can be established between ecosystem conservation status and species richness (Odum *et al.* 1974). Thus, the introduction of exotic species will result in an increase in species richness, without, however, testifying to an improvement in the conservation status of the environment. This increase in diversity is generally transitory and may be followed, in the case of a proliferation of introduced species, by a drastic reduction in biodiversity (banalization of colonizations) (e.g. lionfish in the French West Indies).

Moreover, an optimum of specific diversity seems to be reached when disturbances are neither too rare nor too frequent and when their intensity is not sufficiently strong to cause a global destructuring of the ecosystem (regressive ecological successions). Under these conditions, certain colonizing species, inferred to weak ecological structures, may cohabit with more competitive species found in structured ecosystems. This "intermediate disturbance" hypothesis (Connell 1978) illustrates the difficulty in interpreting a variation in the number of species recorded between two campaigns and the caution necessarily required when using this evaluation criterion.

Another aspect of the compositional description of colonizations is the attribution of an exceptional value (or heritage stake) to certain "remarkable" species. This distinction can be attributed to species because of one or more characteristics that can be linked, for example, to the extractive or non-extractive benefits that they generate (fishing, tourist, scientific, landscape, aquarium, coastal protection, etc.), to their existence and legacy value (traditional value, notions of rarity, endemism, aesthetics, etc.), their interest as bio-indicators (sensitivity to eutrophication, hypersedimentation, haline or temperature anomalies, etc.), their role in the ecosystem (founding species, keystone, engineering, facilitation, symbiotic, etc.) or their degree of threat or regional and international conservation (protected species, ZNIEFF inventories, CITES species, IUCN Red List, etc.).

The intrinsic or contextual characteristics of remarkable species provide keys to the socio-ecological interpretation of fauna inventories. However, these surveys do not take into account the abundance, maturity, size or the state of conservation of the observed colonizations and reflect only an initial exploratory approach to natural environments.

3.4.1.2.2. Community structure

The structural characterization of communities, whether estimated in terms of density, biomass or percentage of cover of the substrate, provides quantitative (or, failing that, semi-quantitative) information that is complementary to the compositional description previously defined.

Indeed, this valuation of the relative contribution of species or groups of species to the community sampled makes it possible to estimate their importance or predominance among the other species or groups of species present. Thus, the dominant or structuring species can represent the "facies" of the biocenosis sampled. This estimate enables a detailed analysis of colonization structures, which, once subjected to the effects of a project, may present certain proliferations or reductions in their numbers without affecting the overall species richness.

As with the compositional description of colonizations, this analysis may be specifically focused on remarkable species (or groups of species), chosen in coherence with each of the objectives of the study (fisheries potential, conservation issue, ecological role, etc.). This approach allows for a more refined ecological interpretation than the compositional description of colonizations, notably because of the range of analyses available for the processing of quantitative ecological information. However, it only gives rare species a low weighting in the analyses, as these species are generally low in number. It is therefore not a substitute for the descriptive approach to colonization, which accords equal weighting to each species surveyed and can thus express the importance of a remarkable species (e.g. keystone species), despite its low representation in the overall colonization.

3.4.1.2.3. Function of biological processes

The functional analysis of ecosystems and biological processes is the final step in the analysis of biocenoses, but the use of which may be of most interest in the eco-design process. It is based on the data acquired during the structural characterization of the communities, by attributing to each species, or group of species (functional group), one or more functional ecological characteristics (diet, demographic strategy, maturity stage, preferred habitat, etc.). These attributes must be derived from reliable and recognized references, which are not always available for the taxa studied. For example, the FishBase[17] site provides, for the study of fish, numerous scientifically validated characteristics, making it possible to update the "species" database acquired during the previous stages of the study of the natural environment and biological balances.

The analysis carried out on the functional groups provides vital information on the possible interactions between organisms (predation, reproduction, etc.), on the demographic structure of the colonizations (maturity, demographic strategies, size classes, reproduction, etc.) and on the main ecological functionalities of the habitats (nursery, growth, rest, protection, concentration sites, etc.).

This information generally enables certain hypotheses to be formed about the possible connections between natural habitats, which may be continuous or fragmented, but above all with regard to the three-dimensional characteristics (design) to be mimicked at the level of the artificial habitats that constitute the infrastructures.

17 www.fishbase.org.

3.4.1.3. *Landscape and connectivity studies*

Once the abiotic and biotic factors have been identified, the analysis of underwater landscape mosaics becomes possible. The landscape is a portion of space represented or observed horizontally as well as vertically by an observer; it therefore implies a point of view. It can have a strong aesthetic dimension, but it covers many meanings and the landscape also manifests the development policies of the territory. In ecology, landscape study may be defined as the study of spatial variations of varying degrees, including biophysical and social causes and consequences of landscape heterogeneity (the concept of landscape ecology), which makes it a necessarily interdisciplinary branch of science.

Even if, during an EIA, the consideration of landscape aspects of a development project is essential, the inventory of underwater landscapes is still an embryonic process and motivates many multidisciplinary studies. However, one phenomenon is of particular interest in order for us to understand the eco-design of structures: landscape connectivity.

3.4.1.3.1. Connections between habitats

Just as in terrestrial ecology, the marine environment is also strongly affected by the partial or total destruction of its habitats, leading to a fragmentation of the ecosystem. In a continuous ecosystem, mobile organisms freely move between habitats (or landscape "patches"). Connectivity, the functional and effective connection necessary for long-term ecosystem functioning, stability and resilience, is normal (Cheminée *et al.* 2014). Species freely move between different points in the underwater landscape, seeking habitats that meet their functional needs, such as feeding, reproduction, resting or, for juveniles, finding trophic (nursery) resources (Figure 3.16).

Nevertheless, if the ecosystem becomes fragmented, with physical barriers or habitat destruction, species move through corridors, connecting two critical habitats, sometimes less favorable for survival (Figure 3.17). This may result in high mortality during movement, which may lead to an interruption in connectivity (Garcia 2014). Remember that habitat alteration, destruction and fragmentation are major causes of species loss and biodiversity reduction, just as their maintenance or development may be an important lever for their increase. The general review of the literature conducted by Bishop *et al.* (2017) demonstrates this.

The impacts of development projects on ecological connectivity within a study area may involve habitat fragmentation (dikes and harbor docks fragmenting nearshore shingle flats), ecological corridor obstruction (pipelines or outfalls across the passage of organisms) and the destruction or degradation of critical habitats (offshore dikes or nearshore roads covering nursery areas).

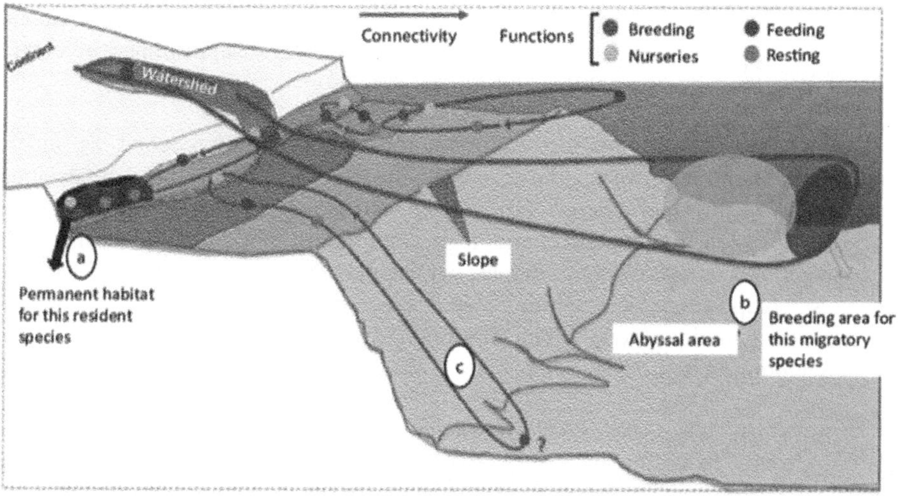

Figure 3.16. *Organization and connectivity of critical habitats for marine organisms (Cheminée* et al. *2014). For a color version of this figure, see www.iste.co.uk/pioch/marine.zip*

a)

b)

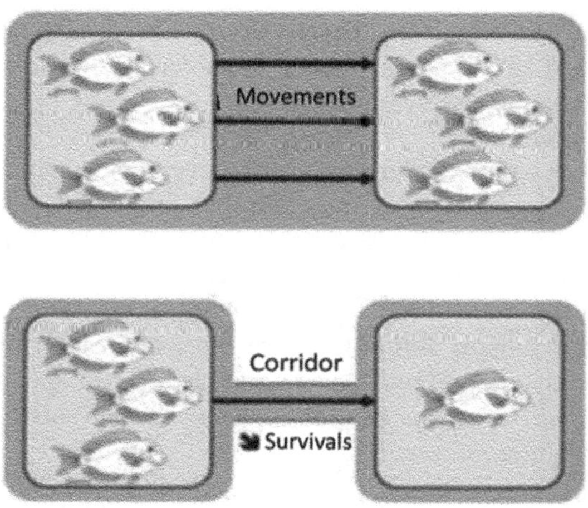

Figure 3.17. *Movement between critical fish habitats: a) continuous ecosystem; b) fragmented ecosystem (from Garcia (2014))*

The consequences on the organisms are generally visible in the medium or long term and are marked by a progressive reduction of certain colonizations and redeployment phenomena in certain reduced species (exploited fishing sites regenerate at varying rates), a structure of colonization characterized by regressive ecological successions (simplifications of the interactions between species and dominance of opportunistic species), even as far as the disappearance of certain species from the landscape of the study area.

Connectivity in the marine environment cannot be approached as in the terrestrial domain. It must take into account elements relating to this three-dimensional environment (Kool and Nichol 2015). The precise identification of ecological corridors (location, functionality) and, *a fortiori*, the estimation of the permeability of these connections (flows) are complex without the implementation of costly sampling protocols (capture–mark–recapture, acoustic telemetry, isotopic analyses, colonization genetics, etc.), which are often disproportionate to the size of the projects submitted to the EIA.

However, in most cases, good knowledge of the environments studied in diving enables experts to define with a relatively sufficient precision, for certain target species (i.e. fisheries in general), the principal habitats, their functions and the connectivities established.

3.4.2. *Integration of the infrastructure into an ecosystem*

Consideration of the ecosystem requires an intellectual approach that integrates many biophysical and landscape parameters. In particular, the concept of "habitat" is key to developing colonizations (Rice 2005).

Compared with ecology, the landscape or biogeographic approach gives primacy to the broad-based configuration of sites, and thus to the general distribution of the habitat, species and movements, whereas an ecological approach emphasizes instead the local interrelationships between species and their immediate habitat (Woillez 2007). When a new infrastructure is built, it creates a new habitat (at the minimum as a hard support substrate), with the colonization of each submerged surface being directly proportional to the surface area of the deployed structure (assuming the support is not a biocide). There are four necessary steps in the eco-design of a structure:

1) definition of habitats and functions created by the infrastructure;

2) listing of target species and their developmental stages;

3) identication of local habitat needs, with respect to target species;

4) design of structures that meet the ethological needs of the target species (functions and habitats).

These steps carried out by the environmental engineer are closely discussed with the civil engineers, so that the two pillars of the design of an infrastructure are aware of the potentialities, and therefore the feasibility, of the project in its ecosystem.

3.4.2.1. *Identification of habitats and functions created by the infrastructure*

Each submerged structure undergoes colonization, unlike land-based structures, and this from the moment of immersion.

At first glance, it would seem futile to claim to be developing "life" on structures, since it develops without any particular effort. However, just as a highway embankment will quickly become green with common and often invasive plants after the asphalt has dried, the species that colonize heavily remediated or anthropized environments are generally ubiquitous, primo-colonizing and of a low level of specialization (Clavel *et al.* 2011).

This homogenization of environments leads to a reduction in sustainable biodiversity, since local species adapted to substrates that meet their functional traits, that is, specificities that make them competitive on "their" natural terrain, are less competitive once the substrates and their habitat are modified (Bulleri and Chapman 2010).

Figure 3.18. *Illustration of the effect of substrate and habitat on colonization. For a color version of this figure, see www.iste.co.uk/pioch/marine.zip*

COMMENT ON FIGURE 3.18. – *On the left, a rockfill dock, a complex habitat colonized (red arrows) by brown and green algae and, underwater (blue arrows), providing shelter and nursery functions for many species of crustaceans, which is extended on the right by a compact and smooth vibrated concrete dock, poorly colonized and with no species underwater (photos and illustration: S. Pioch).*

Invasive species are also often favored by these modified habitats (Bulleri and Chapman 2010). Here, as elsewhere, nature needs attention to thrive, unless we fight against it and deny ecosystems a place on these "human" (or "inhuman" since they reject nature) structures.

However, this vision, in our opinion, is out of step with a proactive development for the environment concerned with biodiversity. Figure 3.18 perfectly illustrates this disposition to welcoming the life that a structure, in this case a harbor dock, may or may not offer, yet with the same functions and the same durability.

The flora (visible on the submerged substrate) and fauna (underwater photos) are diversified and dense as soon as the substrate is adapted to the needs of the species, offering a habitat and interesting functions.

For example, port areas create biotopes[18] that are suitable for certain species, adapted to the habitats and functions. In general, there are three types (Figure 3.19):

1) marine, exposed to the sea, with a fairly agitated body of water;

2) reef, open to the sea, but with a calmer body of water;

3) lagoon, not very open to the sea, with a very calm body of water.

Certain types of habitats and ecological functions are created by these environmental conditions. In addition, these biotopes condition the presence of species at different stages of their development (adults, juveniles), seeking specific abiotic and biotic characteristics (temperature, depth and salinity as well as their prey, such as plankton, algae, mollusks, etc.).

For example, species at an adult stage of development (sexual maturity) will be more present on structures close to the sea, exposed to high agitation conditions (breakwaters or dikes acting as artificial "reefs"), while juveniles will be attracted by low agitation conditions, in sectors with the most sheltered water bodies of the port, whose biotic and abiotic parameters reproduce those of "lagoons" (important natural nurseries) as artificial structures.

18 Environment defined by stable physicochemical characteristics and sheltering a community of living beings (or biocenosis). The biotope and its biocenosis constitute an ecosystem.

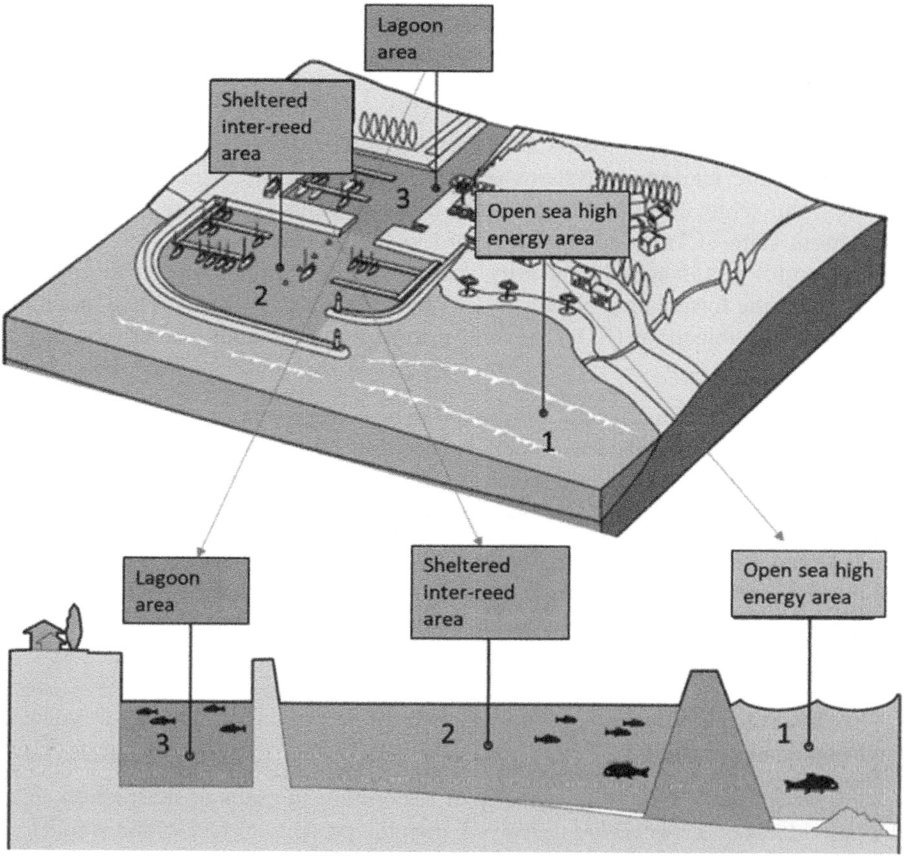

Figure 3.19. *Three harbor areas (in particular, the hydrodynamics), creating three biotopes (biophysical characteristics) favorable to certain habitats, functions and species (S. Pioch). For a color version of this figure, see www.iste.co.uk/pioch/marine.zip*

3.4.2.2. *Selection of target species, habitats and species: prioritization of issues*

As we have seen, submerged engineering structures are frequented by numerous colonizations of fauna and flora, depending, of course, on the physicochemical quality of the environment (pollution, types of substrates, etc.). In order to determine the target species or groups of species, a hierarchy or prioritization must be carried out on the basis of ecological issues[19]. These issues should be considered as elements of marine ecosystems or their functioning, for which good ecological status must be restored or maintained[20].

Among these ecological issues, the following will be considered as priorities[21]: the elements of the marine ecosystems or their functioning for which, depending on the area, the achievement or maintenance of good status is a priority, according to the current state of knowledge with regard to the representativeness of the marine sub-region for this issue, its sensitivity and its functional importance. This approach must enable the formulation of precise (to make them more effective) and ambitious environmental objectives (EOs). Several prioritization methods and criteria can be used:

1) Broad and/or local scale r*epresentativeness*: representativeness refers to the proportion of the issue (species, habitat or ecological function) present in the area under consideration relative to a larger scale. This criterion can be expressed in terms of range, species numbers, total area occupied by a habitat or total biomass.

2) *Large-scale sensitivity or vulnerability*: sensitivity provides information on the intrinsic propensity of the issue under consideration to be destroyed or degraded by pressure and on its ability to recover. Examples:

– intrinsic sensitivity linked to biology (low resistance or resilience);

– limited ecological amplitude which leads to a reduced capacity to adapt to environmental variations;

– rarity which limits the capacity of restoration in cases of destruction (if all the individuals of a species are destroyed, then the capacity of restoration is null).

In many cases, the intrinsic sensitivity of the issue is not known. The choice can then be made on the wider vulnerability of the issue, that is, the risk of seeing the issue disappear or be degraded. This choice is based on the (arguable) assumption that the current level of vulnerability provides indirect information on the sensitivity of the issue to the pressures to which it is (or has been) exposed.

19 The elements presented here are taken from the note "Identification et priorisation de la responsabilité de chaque sous-région marine pour les enjeux ornithologiques", OFB-GISOM, April 2020.

20 For more details, we recommend the *Guide pour l'élaboration des plans de gestion des espaces naturels* (OFB 2018).

21 It should be noted that the local ecological status may be interpreted in contradictory ways to inform decision-making. Thus, the operational choice is sometimes to prioritize actions on issues in good condition in order to conserve them (this is the justification for the naturalness criterion used for the Convention on Biological Diversity) or other times, conversely, to focus on degraded issues in order to restore them.

3) *Functional importance of the ecological issue*: functional importance refers to the determining (or otherwise) character of the issue for the overall functioning of the sector considered or for the species present in the sector, for example:

– an essential sector for the realization of the life cycle of a species;

– a habitat or species determining the functioning of a site (OSPAR criterion, "keystone species").

4) *Additional criteria linked to local specificity*: local specificity combines a set of criteria that enable the issue to be singled out on phenotypical, biogeographical, genetic or phylogenetic grounds, for example:

– a habitat or species with a particular facies not found elsewhere;

– a particular location in the range: range limits, isolated sites;

– local colonization constituting a sub-population of the species;

– a species being the sole representative of a family or genus.

From these methods, it is possible to propose indices that prioritize the issues (which species, habitat, function):

– patrimoniality index: rarity, specificity;

– responsibility index: vulnerability (conservation status, IUCN Red List issues, evolutionary trends), keystone species (engineer or umbrella);

– anthropic pressure index: fishing interest (value, number of professionals targeting the species, etc.), sensitivity to human pressures (pollution, competition, habitat), trends and ecological evolutions or trajectories (colonization monitoring, habitat surface, etc.).

In order to determine the habitat characteristics and natural ecological functions to be reproduced with an infrastructure, it is necessary to identify the target species. Indeed, it would be a waste of time to develop a habitat, with functional characteristics targeting a protection effect, for example, without hoping that one or more species will shelter there.

Species in danger or under strong anthropic pressures, such as corals or marine meadows, exposed to repeated anchorings or to freshwater inputs (emissaries, rainwater canalization), enjoy, in France, a particular protection status which usually prohibits any non-imperative or priority development. Regarding the species without particular status or which are unprotected, we see value in proposing to select the target species of fishing for the design of eco-designed structures. The choice of the ecological issues, which is essential for the definition of the habitats and the functions of the eco-designed structures, may vary according to the patrimoniality or responsibility indices presented above.

3.4.2.3. *Port target species*

As we have seen, submerged engineering structures are frequented by numerous colonizations of fauna and flora, depending on the physicochemical quality of the environment (pollution, types of substrates, etc.).

NOTE.– Port areas must be healthy from a physicochemical point of view (no biocidal chemical substances) and safe from activities presenting a risk of mortality for the targeted species; *otherwise, despite favorable conditions, pressures may modify these orientations*.

In the context of biodiversity erosion, the priority issue is species in danger or under strong anthropic pressure, such as corals or seagrass beds exposed to repeated anchoring or to freshwater inputs (sewage outfalls, rainwater canalization). However, in France, these species have a special protection status, which usually prohibits any non-priority development. As far as non-protected species are concerned, those targeted by fishing should be prioritized for the design of eco-designed structures.

Taking the extension of a port in the Mediterranean area as an example, the principal target species encountered in port and periportal areas (directly linked to an external port structure such as a breakwater, dike, fixed anchor weights or mooring box) are represented in Figure 3.20[22]. Note that a juvenile/adult gradient occurs between the port and the sea, confirming the nursery role of port areas (Bouchoucha 2017).

The species targeted by eco-design that are encountered in ports are generally demersal and benthic (cryptic or non-cryptic).

For example, in the eastern French Mediterranean, these are groupers, black umbers, sars, sea breams, dentex, lobsters, capons, scorpion fish, moray eels, conger eels and pelagics, such as amberjacks, bluefin tuna, dolphinfish, barracudas, pelamids, etc. (Figure 3.21).

The anthropic pressure on certain coastal species allows for a classification of target species of varying priorities based on their selling price, their extreme low numbers or their overexploitation. Clearly, all species are important for the proper functioning of a natural environment, but it is also clear that certain species are under greater pressure, particularly due to fishing.

---

22 Here, we are using the vernacular names; for the scientific names, see the *Guide d'identification des poissons marins* by Patrick Louisy (2002).

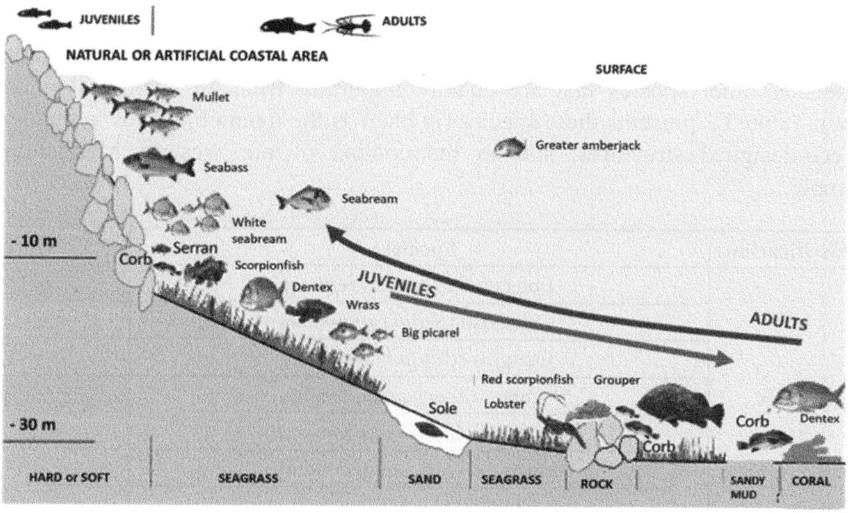

Figure 3.20. *Port and periportal species and their developmental stages in the Mediterranean according to bathymetry and type of seabed (S. Pioch). For a color version of this figure, see www.iste.co.uk/pioch/marine.zip*

Figure 3.21. *Intra-port species in a Mediterranean port from top left to bottom right: school of adult sars (Diplodus sargus), school of post-larval leerfish (*Lichia amia*), juvenile spiny lobster (*Palinurus elephas*) and juvenile seabass (*Dicentrarchus labrax*) (photos: S. Pioch)*

The priority for eco-design is therefore on objectives adapted to these target species, in order to generate habitats and functions that meet their needs (without neglecting other species that are equally important from an ecological point of view). Table 3.2 presents these species (in blue) as the main objectives in the design of eco-designed structures, still in the context of our western Mediterranean example.

Classification	Species	Pressure
Benthic	Lobster (*Palinurus elephas*)	+++
	Cuttlefish (*Sepia officinalis*)	-
	Octopus (*Octopus vulgaris*)	++
	Capon (*Scorpaena scrofa*)	+++
	Scorpion fish (*Scorpaena notata*)	+
	Moray eel (*Muraena helena*)	-
	Conger (*Conger conger*)	-
Demersal	Red mullet (*Mullus surmuletus*)	++
	Mullet (gray, golden gray and thinlip) (*Chelon labrosus, Liza aurata* and *Liza ramada*)	-
	Squid (*Loligo vulgaris*)	++
	Seabream (*Pagellus erythrinus* and *Pagellus acarne*)	+
	Sar[23] (*Diplodus sargus*)	++
	Grouper (*Epinephelus marginatus*)	Protection
	Corb (*Sciaena umbra*)	Protection
	Bass (*Dicentrarchus labrax*)	+++
	Sea bream (gilthead and black) (*Sparus aurata* and *Spondyliosoma cantharus*)	++
	Dentex (*Dentex dentex*)	+++
	Red porgy (*Pagrus pagrus*)	++
Pelagic	Barracuda (*Sphyraena viridensis*)	-
	Mackerel (*Scomber scombrus*)	-
	Yellowtail (*Seriola dumerili*)	+++

Table 3.2. *List of target species (benthic, demersal and pelagic) in port and periportal areas and the level of pressure experienced in the framework of a project in the eastern French Mediterranean area (very low -, low +, medium ++/orange and high +++/red). For a color version of this table, see www.iste.co.uk/pioch/marine.zip*

23 In the sar family, only *Diplodus sargus* is of high fishing importance, while *D. annularis, D. vulgaris, D. puntazo and D. cervinus* are under less pressure (no fishing or economic data).

The highest pressures, symbolized by "+++" and "++" indictors, enable the identification of the following *target species* (vernacular name followed by the scientific name): spiny lobster (*Palinurus elephas*), octopus (*Octopus vulgaris*), capon (*Scorpaena scrofa*), red mullet (*Mullus surmuletus*), squid (*Loligo vulgaris*), sar (*Diplodus vulgaris*), bass (*Dicentrarchus labrax*), gilt-head bream (*Sparus aurata*) and black seabream (*Spondyliosoma cantharus*), dentex (*Dentex dentex*), red porgy (*Pagrus pagrus*) and yellowtail (*Seriola dumerili*). The grouper (*Epinephelus marginatus*) and the corb (*Sciaena umbra*) are under special protection, so we include them in the priority list.

Target pelagic species such as bluefin tuna, swordfish, pelamids or dolphinfish may be present in the coastal port areas of the eastern Mediterranean. However, this presence is exceptional, except for yellowtails and barracudas, since their preferential area of evolution remains the open sea. Strictly, pelagic species such as tunas, swordfish, dolphinfish or pelamids should therefore not be targeted.

The target species that can guide the functional design of eco-designed structures, in the case of a port in the Mediterranean area, can thus be divided into three types: (i) benthic: spiny lobster, capon, octopus; (ii) demersal: red mullet, seabream, grouper, corb, sea bass, black and gilt-head bream, dentex, porgy, squid; (iii) pelagic: yellowtail. As a reminder, for these target species, we recommend favoring the installation of associated sessile or vagile species specific to shallow coastal waters.

3.4.2.4. *Targeted stages of development*

Submerged infrastructure creates new habitats and ecological functions that are suitable for certain local species of fish, crustaceans and mollusks at different stages (juvenile/adult) of their development or different periods and seasons.

Consequently, eco-designed equipment may be arranged and sized to favor its role as habitat and function, adapted to sessile (fixed: algae, mollusks, etc.) or vagile (mobile: fish, crustaceans, etc.) species, at juvenile or adult stages, and depending on certain periods or conditions favorable to its use.

For example, in the case of demersal and benthic target species, in order to optimize the positioning and shape of infrastructures, it would be necessary to show the habitat requirements adapted to certain developmental stages. Depending on the stage – adult or juvenile – targeted, the preferential conditions of agitation sought are different; adults will seek more agitated areas while juveniles will seek calmer conditions. These functional needs influence the selection of sites suitable for installation of structures, on the one hand, and suitable for types of favorable habitat (large habitats for adults, smaller for juveniles), on the other. The cross-referencing of these data makes it possible, such as in the case of a port project, to ensure that the same structures are not installed in different basins and sites (based on the line of thought in Table 3.3).

Habitat for juveniles	**Habitat for adults**	**Preferred conditions**	**Port installation sites**	**Favorable habitat type**
	X	Exposed	1 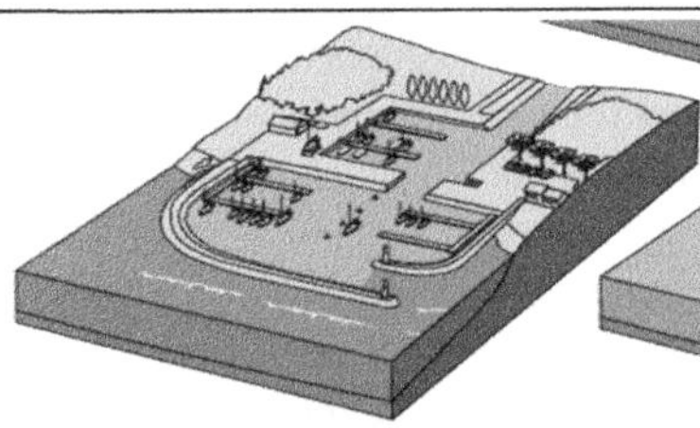	Macro-habitats
X	X	Mildly agitated to calm	2 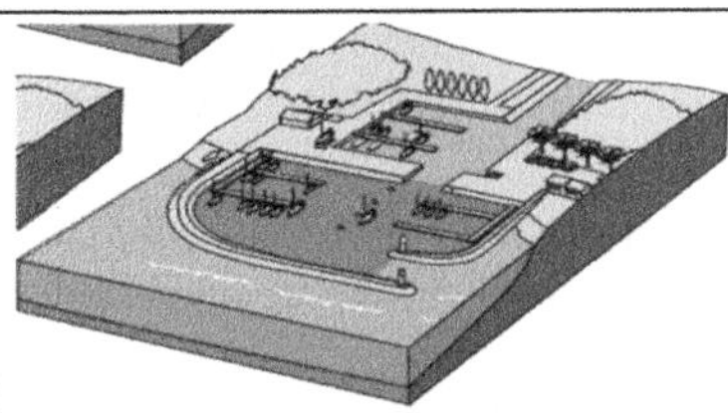	Micro- and macro-habitats
X		Very calm	3 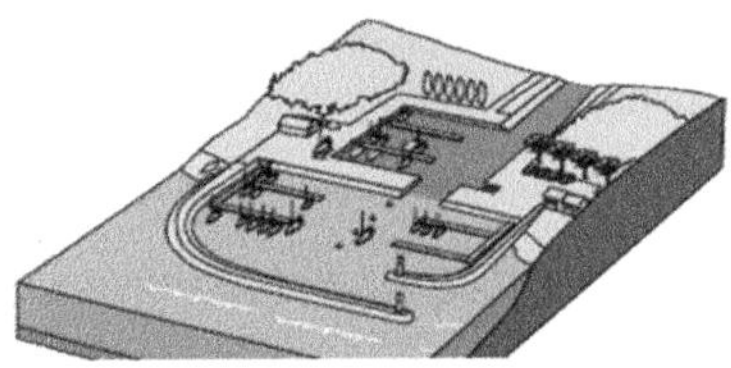	Micro-habitats

Table 3.3. *Preferred agitation conditions vary according to the adult or juvenile stage targeted, which points to a different choice of harbor installation site and favorable habitat type within the same development. For a color version of this table, see www.iste.co.uk/pioch/marine.zip*

3.4.2.5. *Habitat requirements and functions of target species*

Research, which enables us to define the vital needs of the species (essential habitats, functions, biotic and abiotic conditions of the environment, ecological interactions), has a direct influence on their growth and survival (see in particular Kuroki (1952)).

In order to define suitable habitats for target species, they are classified according to their ethological group and habitats, as proposed by Nakamura in 1985 in his work on artificial habitats created by human for target species (Figure 3.22).

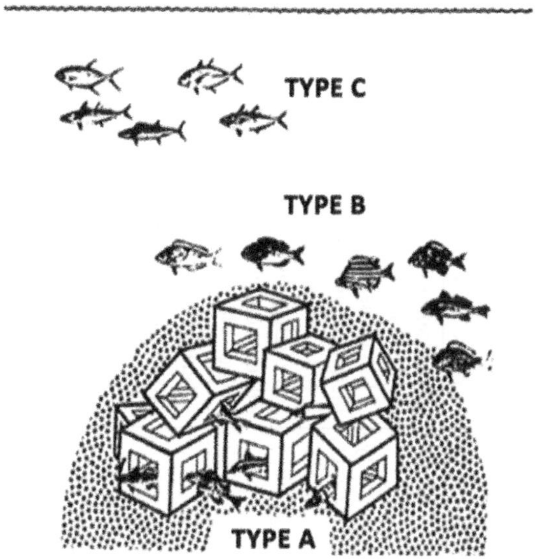

Figure 3.22. *Types of target species based on their relationship with artificial habitat A, B or C (from Nakamura (1985))*

In order to facilitate the design of eco-designed structures, we propose to use this approach, where a typology of species is defined from an ethological point of view according to three homogeneous groups called "*types of target species*":

– *type A (or I): benthic* or cryptic bottom-dwelling species whose natural habitat is a hard substrate, generally with dark and complex cavities, and with which they are in contact;

– *type B (or II): demersal* and coastal species living on or close to the seabed, whose natural habitat is a hard or soft substrate presenting, in the case of a hard substrate, a relief and one or more medium to large cavities. These species remain in visual or sound contact with hard substrates and near the seabed;

– *type C (or III): pelagic* species living in open water[24] and seeking the proximity of steep upwellings on the seabed or presenting a very strong relief (creating upwelling[25] or micro-upwelling currents).

Species are classified through observation of their behaviors, as it determines their relationships with artificial (or natural) habitats.

24 Generally from the surface to half the height of the water column, according to Nakamura (1985).

25 Upwelling: ascending currents of deep cold water rich in nutrients.

Types of target species	Behaviors	Natural habitats sought	Adapted AH categories	Complexity of the eco-designed module			Species requirement
A	Benthic Low displacement Local scale	Hard substrates with dark and complex cavities	1	Numerous and complex shelters	Low cavity height $\geq$ target species	Internal spaces close to the size of benthic species (narrow, often cryptic cavities)	Majority or part of the body in contact with the artificial habitat
B	Demersal Medium to large displacement Regional and national scale	Hard substrates with medium to large cavities	2	Low complexity shelters	Average cavity height > target species height and width	Internal spaces large enough for demersal fish to visually identify themselves fully and move through cavities without coming into contact with the AH	Little physical contact with the artificial habitat but swimming in the vicinity
C	Pelagic Major migratory movements International scale	Open water and the proximity of steep or very steep upwelling	3	No shelters	Greater or lesser cavity height	Shape creating current disturbances and/or a position between the surface and the middle of the water column above the mean thermocline	Away from the AH and living in open water

Table 3.4. *Relationship between target species types and eco-design (S. Pioch)*

The shape, or "design", of submerged structures must therefore take into account these ethological characteristics. Certain rules may be laid down in order to guide the civil engineer in the shape of structures on the basis of the recommendations and knowledge of the environmental engineer.

Table 3.4 enables the definition of structure shapes that are suitable for the needs of these three categories.

The main ecological (structural and functional) or landscape functions pursued by eco-designed structures are generally of three kinds:

– habitat: nursery, protection, spawning ground;

– trophic: feeding with grazing (algae/mollusks) and predation (carnivores);

– landscape connectivity: for example, maintaining a coastline, improving connectivity in hard substrates, corridors between seagrass beds.

The environmental conditions and areas occupied to provide these different functions may vary (depth, substrate, temperature, etc.) and evolve through the different developmental stages, from juvenile to adult (Cheminée *et al.* 2014).

Other ecological functions provided by the species may exist. Examples include:

– filtration;

– sediment fixation;

– bioconstruction;

– mineralization of organic matter.

However, these functions may be considered secondary to the objectives of an eco-designed project and the ecology of the target species. To continue with our harbor example, they would be primarily dedicated to the nursery, feeding and connectivity (larvae/juvenile) functions described above.

These essential or basic functions (reproduction, nurseries, nutrition, protection, connectivity) are linked to specific habitats (design, depth, luminosity, orientation with currents, substrates, etc.) for the listed target species. The success of an eco-designed project consists of advocating a structure design that specifies the function(s) favored and for which species, at which stage(s) and in which harbor sector(s).

3.4.3. *Bio-inspiration and design of eco-designed structures*

We already touched on this subject when we discussed the concept of working with nature, but we should remember that this is one of the most important objectives of an eco-design approach: to be inspired by a natural environment in all its ecological and aesthetic dimensions.

Indeed, seeking biomimicry is an interdisciplinary conceptual approach, taking nature as a model to meet the challenges of sustainable development (social, environmental and economic). Moreover, it is not out of the question that, in certain cases, the technical answer cannot be inspired by the example of nature.

The eco-design of port structures is part of this objective, particularly landscape integration in order to maintain structural continuity (landscape connectivity) with the surrounding hard substrates. In 2017, an installation of biomimetic substrates was carried out in the Gulf of Ajaccio, under the aegis of the Corsican environmental office (Figure 4.37).

3.5. Eco-design at the material level: the example of concrete[26]

3.5.1. *A little history*

The study of materials used may seem, at first sight, to be a technical detail; however, it is revealing of the history and philosophies of marine development. It demonstrates the investments made by countries to develop this tool in terms of financial investments as well as in terms of reflection on their implementation. In some cases, the use of materials from industrial waste contributes to accentuating the negative image for developments inserted into a natural environment: the sea.

The materials used must meet criteria of "functionality", compatibility, sustainability, stability and availability under the specific and very "aggressive" conditions of the marine environment (abrasion, corrosion, stress, etc.) (Lukens and Selberg 2004). This history is more than 400 years old and dates back to the time of Emperor Jôo in Japan in the 17th century, from when the first written records of the design of fishing structures from blocks of rock placed in shipwrecks submerged in the sea were found. The technical description is very precise, and the material aspect is linked from the outset to questions of habitat (volume, ecological functions) and durability (ships' timber in particular).

---

26 This section was co-written by Jean-Claude Souche, Sylvain Pioch and Marie Salgues, assistant professor at IMT Mines Alès.

However, it was not until the 20th century that this structural diversity was truly "exploited". Table 3.5 summarizes the different materials listed in two artificial reef design guides (Seaman and Sprague 1991; Fourrier *et al.* 2015) and their primary uses.

Although it appears to be a very large choice *a priori*, it is advisable to act with caution. In fact, many experiments involving the immersion of salvaged materials (tires, carcasses, wrecks, etc.) have been carried out in the past in the United States with sometimes mixed or catastrophic (in the case of tires) results.

In France, the environmental administration (DREAL) vigilantly ensures that products that could be assimilated to waste are not implemented as artificial reefs, in compliance with the OSPAR Convention (Convention for the Protection of the Marine Environment of the North-East Atlantic, adopted in 1992). These structures must be designed in relation to their use and be aesthetically pleasing (Fourrier *et al.* 2015). These considerations greatly reduce the choice of artificial materials listed in Table 3.5.

Today, new approaches aim to encourage all of the actors (owners, engineers, contractors) in construction to use alternative materials or concepts for concrete structures whenever possible – and why not those inspired by nature? For example, the presence of mangroves has a beneficial effect in absorbing tsunami waves as well as typhoon swells in Asia[27]. It goes without saying that such solutions must be put into perspective within the framework of the complete development of a coastline and the vision of the land that is being created.

However, in many cases, when artificial structures made from construction materials such as concrete are to be used, increased structural complexity is required to support biodiversity, ecosystems and the services they provide (Cooke *et al.* 2020).

The choice of the companies has mainly been based on the manufactured materials mostly used in construction: concrete and steel. These inert materials require a transformation that consumes a lot of energy to make them usable on a construction site. Once in the natural environment, they will slowly degrade over several decades to return to their initial "calcite" or "iron oxide" state.

To promote strength and durability (50 years at least), the concrete generally used is a compact concrete which has been adapted to the marine environment (AFNOR 2014). Concrete possesses a good compressive strength, and its fresh state plasticity associated with the use of molds enables the manufacture of elements of various shapes in series at moderate cost. On the other hand, its tensile strength is low, which may require the use of internal steel reinforcements.

27 https://www.publicplanetpartnerships.com/2019/11/05/tsunami-proof-mangrove-seawalls/.

Natural materials	Main uses	Artificial materials	Main uses
Vegetable fibers: Bamboo Coconut fibers Biosourced ropes	Fisheries	Recovered materials: Concrete (demolitions) Slag Tires Car bodies, tanks Decommissioned oil platforms Tramway, wagons Vessels (wrecks)	Recreational, experimental Recreational, fishery Recreational, experimental fishery Recreational Fisheries Recreational, experimental Fisheries
Mineral materials: Compacted shells Rock boulders, gabion rocks Wood (pruning wood, wood planks)	Experimental, compensation Recreational, experimental, compensatory measures Experimental, recreational	Manufactured materials: Fiberglass, polymers Metal Prefabricated concrete	Recreational, fishery Recreational, experimental fishery All uses

Table 3.5. *List of different materials used, excluding waste, in the manufacture of artificial reefs and usage terminology as indicated in the submersion projects (adapted from Seaman and Sprague (1991))*

Materials	Advantages	Disadvantages	Comments
Concrete	Easy to shape material Surface suitable for organism fixation, Strong and durable (50 years) Non-polluting Low cost	Handling and transport (weight) Weakness of concrete in tropical areas (temperature and salinity)	Most widely used material Numerous studies available and positive feedback Wide variety of compositions and textures Easy to create holes and cavities
Steel	Good mechanical qualities Wide range of architectural possibilities Interesting resistance for large structures	Oxidation and corrosion heterogeneity Assembly of parts (welding, rivets, bolts, etc.) Higher cost Durability of 20–30 years	In Japan, metals are sometimes protected by by concrete Aesthetic aspect of the structures after a few years can be improved Electroplating can be used to to "naturalize" the surface with natural carbonates (http://www.geocorail.com/; Hilbertz 1978)
Stone	Good fixing substrate Good mechanical and chemical resistance to marine conditions Natural landscape aspect Low cost	Low architectural possibilities Not easy to handle No intrastructural cavities and reduced developed surface	"Natural" aspect enhancement Aesthetic and landscape integration Piles of rockfill resembling natural scree
Plastic	Good resistance to marine conditions Great architectural possibilities Flotation capabilities	Cost in general very high Pollution perception if degradation Low density (anchors required) Smooth surface, not very suitable for colonization	Its use is often limited to the construction of mobile devices for shellfish farming, floating devices or the assembly of modules of another material

Table 3.6. *Summary of the advantages and disadvantages of the main materials used for artificial reef construction (Lukens and Selberg 2004)*

After concrete, metal is the most widely used. It is an easy-to-use material which, by assembling it in sheets, can adopt almost any shape imaginable. Steel possesses a high elastic resistance, which allows it to support strong traction and bending without damage. It is particularly interesting in the formation of frames or structural frameworks. Its main weakness in a marine context is corrosion, which conditions its life span, usually around 20–30 years. The installation of sacrificial anodes may extend the life span and should be considered, given the potential release of heavy metals.

It should be noted, however, that the comparison of these two materials shows that structural possibilities are complementary and that their combination offers a wide range of possibilities. Table 3.6 compares the advantages and disadvantages of some of the materials (concrete, steel, natural stone, "plastics") usually used in the construction of artificial reefs.

Finally, concrete, which is widely used, is recognized as resistant, durable and favorable to the implementation of biodiversity. Indeed, once submerged, it undergoes a spontaneous colonization by opportunistic, ubiquitous, primo-colonizing organisms: bacterial biofilm followed by unicellular algae and then, with time, more complex organisms (bivalves, tube worms, tunicates, hydroids, etc.). The cycle and the speed of colonization will depend on the environmental conditions, as well as on the nature of the substrate (composition, physicochemistry of the support, etc.).

Before detailing the construction material concrete, the most consumed worldwide and the second most consumed with water, it seems appropriate to note the opinion of industrialists specialized in this material, and their interest with regard to the eco-design approach, through the words of Patrick Guiraud.

"Ecological inventories reveal an accelerated decline and erosion of biodiversity over the last several decades, resulting in a degradation of natural environments and ecosystem services offered by nature and a destruction of ecological continuities. The main factors of biodiversity loss are clearly identified: reduction and fragmentation of natural habitats, overexploitation of natural resources, air, water and soil pollution, growth of urbanization, infrastructure development and the effects of climate change.

The preservation and restoration of biodiversity is therefore now a major societal challenge and recognized as being as important for the future as the climate issue. It is now essential to aim for better conservation of favorable, species-rich habitats, to improve the movement of species and to restore and safeguard this natural wealth and to adopt an offensive attitude of creating biological wealth by taking inspiration from nature.

– Biodiversity, our natural wealth: biodiversity refers to the diversity of all forms of life. It includes the diversity of genes, species and ecosystems and all the relationships and interactions between living beings themselves and their living environments. It thus concerns all of the natural processes which ensure the perpetuation of life in all its forms. It is a saving inherited from the past, the fruit of 4 billion years of evolution. Biodiversity provides ecosystem services that are essential to human well-being and to the economic, social and cultural development of our societies and which contribute to the regulation of the planet's equilibrium and to mitigating the effects of climate change.

– The advantages of concrete for biodiversity: concrete is an excellent medium for preserving, restoring and creating biodiversity thanks to its multiple properties.

Its neutrality (no ecosystem or water disturbances), durability, resilience and plasticity (enabling the creation of structures favorable to wildlife) are all potentialities for creating natural habitats, diversified reception areas, shelters for wildlife, etc.

Concrete is a base on which life can cling, a support for life and the development of flora (mosses, lichens, plants, etc.) and a refuge for biodiversity. Concrete, with its recognized qualities, has all the arguments to take up the challenge of preserving and restoring biodiversity, particularly in marine sites.

– The preservation of biodiversity in marine sites, the example of artificial reefs: the fight against the destruction of natural coastal habitats is a new challenge, relating to the restoration or creation of living spaces for biodiversity by installing artificial reefs, which enable the development and reproduction of different species constituting the marine ecosystem. Each artificial reef requires a tailor-made response, adapted to local conditions and the objectives pursued. A wide variety of artificial habitat modules are made of concrete, which offers great flexibility of use and quality of reception of biodiversity.

The eco-design of these structures is experiencing interesting developments, with each seemingly an opportunity to demonstrate the ecological qualities of concrete at sea. Concrete favors biological colonization through the variety and potential complexity of its forms, surface conditions and roughness. The Ricantu Reefs operation in Corsica is an exemplary illustration of this.

Research is also underway to improve the biological colonization of all marine structures, to make concrete 'biogenic' and 'biomimetic', drawing inspiration from nature and models that have been working for billions of years to create even more efficient and virtuous habitats. We must reconcile concrete and biodiversity, and draw inspiration from nature to innovate.

The concrete industry therefore supports this logic of progress and research initiatives that aim to safeguard, preserve and develop positive biodiversity and reconcile the future of humans with that of nature. In particular, it supports approaches and innovations inspired by the intelligence of living organisms with a systemic approach to the service of humanity and life.

By working together, the ecological engineering and concrete industries will find concrete solutions to restoring biodiversity to its rightful place on our planet, and in particular in our oceans, with an innovative approach based on living organisms and in harmony with nature."

Box 3.6. *What is the vision of the concrete industry with regard to eco-design? (source: Patrick Guiraud[28])*

3.5.2. *Concrete, the state of the art*

3.5.2.1. *Definition and constituents of concretes*

Concrete is a mixture (Figure 3.23) defined by a formulation. It must meet performance requirements, such as consistency in its fresh state and mechanical strength in its hardened state. From the choice of constituents, its expected properties are obtained through an adapted manufacturing protocol (Charonnat 2012).

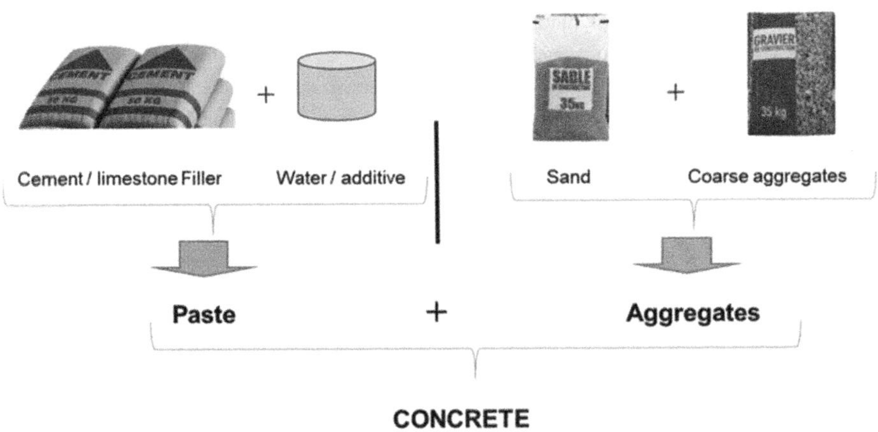

Figure 3.23. *Summary diagram of the constituents of concrete (J.-C. Souche)*

Concrete varies according to the nature of the component aggregates, admixtures, colorants and surface treatments, and can therefore be adapted to the requirements of each project in terms of its performance and appearance.

28 Civil Engineering Director of the French professional organization CIMbéton.

The choice of the constituents is essential; in particular, the particle size distribution of the various aggregates, which, in turn, depends on the objectives to be reached in terms of granular stacking of the concrete (De Larrard 2000; Festa and Dreux 2007). Paste is an active element of the concrete; it coats the aggregates and fills the voids existing between the grains. The paste acts as a lubricant in the fresh state and as an adhesive in the hardened state.

The dosage of water is therefore a key parameter (Festa and Dreux 2007) since it conditions the hydration of the cement, the workability of concrete in the fresh state and the achievement of compressive strengths in the hardened state at 28 days (fc_{28}) (Bernier 2004). The choice of the type of cement and its dosage depend on the performances required (mechanical resistance, resistance to aggressive agents) and, at the same time, on the nature of the other components. The cement dosage depends on several criteria such as the type of concrete, the destination of the structure, the required strength and the environmental actions to which the concrete is subjected.

There are several formulation methods, including the Bolomey, Féret and Dreux-Gorisse methods, modified by the MPT[29] (De Larrard 2000), which explain fc_{28} as a function of the quantity of effective or total water (Festa and Dreux 2007), expressed in volume (Ve), mass (W_{eff}[30]) or through the water/cement mass ratio (W_{eff}/C) (Bernier 2004; Yamine 2007).

The minimum W_{eff}/C[31] dosage to be expected corresponds to the amount of water that will just fill the void volume of the granular stack (De Larrard 2000). For a given W/C, the concrete with the most compacted granular pile (optimum granularity) will have the best workability in the fresh state (Yamine 2007).

The mechanical properties of concrete and the transfer properties (porosity, permeability) are dependent on the W/C ratio (Salgues 2013). Increasing the amount of water beyond the minimum dosage will control the workability of fresh concrete and increase the porosity of hardened concrete and thus influence its durability (De Larrard 2000; Bernier 2004).

29 Maximum paste thickness (De Larrard 2000).

30 "The term Effective Water (W_{eff}) refers to the portion of the total water in a m^3 of concrete available for cement hydration. It is also the total water less the water absorbed by the aggregates. In fact, the water absorbed by the aggregates does not participate in the workability, nor in the hydration" (Bernier 2004).

31 For hardened concrete, the application of Powers' hydration model shows W/C values in the range of 0.39–0.42 (Powers and Bromnyard 1946; Jensen and Hansen 2001).

3.5.2.2. *Colonization of concrete in the marine environment*

Concrete structures in contact with the marine environment modify the natural environment and subsequently become new ecosystems whose settlement mechanisms must be understood (Bulleri and Chapman 2010). However, this understanding is hampered by the many parameters involved: on the one hand, environmental factors (Table 3.7) and, on the other hand, those related to the substrate itself (Table 3.8). The combination and interactions between these different parameters will drive the colonization of the structures.

Conditions	Effect	Location/duration/reference
Immersion period	Different assembly depending on the immersion period	Australia/1 year (Anderson and Underwood 1994)
Temperature	Principal factors concerning the colonization of ciliates: temperature, salinity, nutrients	China/1 year (Gong *et al.* 2005)
Brightness	Existence of an optimum according to the species	Australia/4 months (Mundy and Babcock 1998)
Ecosystem	The same structure in different locations will generate a different assembly	Sherman *et al.* 2001
Hydrodynamics	Importance of hydrodynamic conditions on colonization	Sheng 2000

Table 3.7. *Effect of environmental conditions on marine colonization*

Biofilm colonization and development on a cementitious surface have been the subject of several studies showing that biocolonization of a surface is induced and governed by the interaction of three actors: the material, the living microorganisms and the environment in which the first two are found (Lors *et al.* 2017).

For this purpose, two types of bioreceptivity are defined: intrinsic bioreceptivity, related solely to the physicochemical properties of the concrete surface (porosity, rugosity, etc.) and the chemical composition of the material, and extrinsic bioreceptivity, related to the environment (Lors *et al.* 2017).

Many works clearly indicate that the nature of the material affects the kinetics of biocolonization, the distribution of adherent bacterial cells and the formation of the biofilm. The differences observed between materials may be induced by their chemical natures but also more importantly by their physicochemical properties or topography (Guillitte and Dreesen 1995; Bellon-Fontaine *et al.* 2016). The physicochemical properties of a cementitious material may therefore influence bioreceptivity directly or indirectly, by acting on the other factors influencing bioreceptivity.

Substrate	Effect	Location/duration/reference
Composition		
Concrete Plywood Fiberglass Aluminum	Different blend after one or two months but tending to become similar after 4–5 months	Australia/1 year (Anderson *et al.* 1994)
Concrete Wood Steel	Little effect of the nature of the substrate on the blend and the degree of colonization	Japan/18 months (Choi *et al.* 2006)
Limestone Basalt	Same blend but different quantity	Australia/10 months (Green *et al.* 2012)
Texture		
Concrete, smooth surface Limestone, smooth surface Granite, rough surface	Different blend	England/8 months (Coombes *et al.* 2011)
Concrete, different texture	Textured concrete improves colonization and growth	Netherlands/18 months (Borsje *et al.* 2011)
Coating with chemoattractant elements	Improves colonization and growth	South Korea/90 days (Lee *et al.* 2008)
Design		
Slope of the support	Decrease of biomass based on angle, different blend	Japan/7 weeks (Somsueb *et al.* 2001)
Presence of a shelter	Refuge from predators	Panama/8 months (Menge *et al.* 1983)
Durability Concrete, limestone, granite	Influence of the blend on the durability of the substrate	England/32 months (Coombes *et al.* 2017)

Table 3.8. *Effect of substrate on marine colonization*

Another advantage of concretes is related to their ability to recycle materials such as recycled aggregates (Recybeton 2012), hemp- or oyster shell-based aggregates (Dennis *et al.* 2018), or fly ash or blast furnace slag (McManus *et al.* 2018). Recycling improves the environmental footprint of structures, in particular the carbon footprint. Some authors go as far as proposing the reuse of sections of structures in new works (Lacovidou and Purnell 2016). All these approaches are potentially interesting and can be considered for marine structures, since they help reduce the environmental impact of structures and because they can promote colonization by living organisms. Various authors have demonstrated that "low pH" concrete based on CEM III is favorable to colonization by biofilms or macrophytes (Perkol-Finkel and Sella 2014; Hayek *et al.* 2020), or that the carbonation of concrete, which significantly reduces the pH of the surface of the material, also has a positive effect on the installation of biofilms (Hayek *et al.* 2020). However, it should be kept in mind that structures at sea are subject to multiple physical and chemical stresses. Consequently, anything that is conducive to colonization should not be detrimental to sustainability (Souche *et al.* 2016). Finally, the use of industrial byproducts could also promote the release of trace amounts of toxicants, such as heavy metals (McManus *et al.* 2018), and should be checked beforehand.

It thus appears that the scope offered by concretes is vast. The solutions chosen must reconcile the strength and durability of the structures while encouraging colonization by living organisms, and native species in particular, which must not be disadvantaged by invasive species. Under these conditions, concrete structures can be used in a positive way to fight against the effects of global warming.

3.5.2.2.1. Chemical composition

Depending on its chemical composition, a material may contain substances that promote or inhibit the development of microorganisms. For example, the presence of certain heavy metals can inhibit their growth and development (Dubosc 2000; Dalod 2015).

In addition, the chemical and mineralogical composition of cement can influence other material parameters, such as roughness, porosity and pH (Lors *et al.* 2017).

3.5.2.2.2. Surface roughness

The roughness or surface topography of materials is one of the main physical factors influencing its bioreceptivity. Roughness translates into asperities and irregularities on the surface of the material. These irregularities can form anchorage sites and micro-refuges for microorganisms to settle, thus promoting biocolonization (Ammar *et al.* 2015).

These microrefuges protect microorganisms against hydrodynamic forces that tend to pull them out, especially during reversible, highly fragile adhesion (Kerr *et al.* 1999; Bellon-Fontaine *et al.* 2016). Researchers have observed an increase in bacterial adhesion on glass and acrylic samples immersed in natural seawater when the roughness increased from 0.25 to 30 μm (Kerr *et al.* 1999). As an example, Teughels *et al.* (2006) showed that bacterial plaque formation is favored on a rough surface.

Regarding cementitious materials, several studies have demonstrated the effect of surface roughness on biofouling (Dubosc 2000; Tomaselli *et al.* 2000; Miller *et al.* 2009; Tran *et al.* 2012, 2014; Govin *et al.* 2013; Manso *et al.* 2014; Ahmed 2016; Manso and Aguado 2016). These studies have shown that rough, cementitious surfaces have a higher bioreceptivity than smooth surfaces under controlled laboratory or natural conditions.

In addition to its effect on the adhesion of microorganisms, surface roughness also influences the water flow velocity as well as the binding of organic matter (Dubosc *et al.* 2001; El-Naggar *et al.* 2010). This directly impacts the formation of the conditioning biofilm at the origin of the biofouling process.

Some studies qualify the effect of roughness on the bioreceptivity of materials by stating that this effect depends on the characteristics of the microorganisms in question (type of microorganism and its geometry, size, shape, etc.) (Whitman *et al.* 1998). For example, filamentous algae, such as *Klebsormidium flaccidum*, cling to rough surfaces, while thick-walled algae adhere to smooth surfaces (Lors *et al.* 2017).

3.5.2.2.3. Porosity

Bacteria preferentially colonize porous supports. This finding is mainly explained by the influence of porosity on the absorption and retention of water by the material determining the development of microorganisms (Dubosc 2000; Nuhoglu *et al.* 2006; Guézennec 2017).

In the case of immersed concrete, it is difficult to determine the effect of porosity of a cementitious material on bioreceptivity (Tran *et al.* 2011, 2014; Govin *et al.* 2013). This difficulty can be explained by the permanent saturation of the samples tested with water and thus the high humidity throughout the experiment. However, the porosity of a cementitious material remains an important parameter in its durability. This translates into a potentially faster deterioration than that of a less porous material (Fritz-Feugeas *et al.* 2008; Drouet 2010).

3.5.2.2.4. Surface pH

The pH of the material surface may define the nature and ability of a microorganism to attach to it (Shi and Zhu 2009). Generally, each species has an optimal pH for growth and can grow within its own defined range around this pH. According to these ranges, microorganisms are classified as neutrophilic (pH between 6 and 8), acidophilic (pH below 6) or basophilic (pH above 8).

Microorganisms' metabolic activities, growth, adhesion and EPS synthesis are influenced by the pH at the material surface (Roux *et al.* 2006; Ahmed 2016). Reducing the surface pH of a concrete after immersion in seawater (<9) accelerates surface biocolonization and improves the bioreceptivity of the material (Vincke *et al.* 2001; Roberts *et al.* 2002; Tran *et al.* 2012; Dalod 2015; Ahmed 2016).

Moreover, extremophilic bacteria can resist extreme environmental conditions such as temperature, pH and pressure and can colonize concrete with a very basic surface. It remains to be seen whether these bacteria are the right targets for the installation of endemic, rather than invasive, species.

Although substrates based on cementitious materials have a disadvantage related to their basic pH that would seem unfavorable for the development of biota (Perkol-Finkel and Sella 2014), they are usually used as a solution for the design of eco-designed structures. In particular, concretes based on CEM III or CEM V cement see lower pH values (of the order of 11) than concretes based on artificial portland cement (CEM I or CEM II). It should be noted that the surface pH evolves rapidly over time when the concrete is immersed in sea water. Therefore, this handicap seems to fade rapidly over time (De Weerdt *et al.* 2014; Perkol-Finkel and Sella 2014).

3.5.2.2.5. Surface hydrophobicity

Hydrophilic microorganisms bind more easily to hydrophilic surfaces, while hydrophobic microorganisms prefer hydrophobic surfaces (Shi and Zhu 2009). Note that the hydrophobicity and charge of a surface depend on the chemical composition of the material (Katsikogianni and Missirlis 2004). According to several studies, this composition influences the type and diversity of colonizing microorganisms (Guillitte 1995; Giannantonio *et al.* 2009; Miller *et al.* 2012; Dalod 2015).

3.5.3. *Concrete formulations*

3.5.3.1. *The standard approach*

For the material concrete, the standard NF EN 206/CN 2014 (AFNOR 2014) defines a prescriptive approach based on an obligation of means relating to the resistance and the durability of the concrete. The requirements of this standard depend on the aggressiveness of the environment of the structure and its duration of use. It provides the minimum dosages to be respected according to the concrete exposure classes, according to the environmental actions (moisture, marine environment, chemical aggressions, freeze–thaw cycles).

The standard NF EN 206/CN constitutes the basic text for prescribing, formulating, manufacturing and controlling concretes. It is part of a coherent set of standards. It defines requirements for the durability of structures, precisely takes into account the concept of exposure classes, imposes strict requirements on manufacturing controls, specifies the distribution of responsibilities between the prescriber, the producer and the user and the role of each actor and defines provisions relating to conformity tests, assessments and criteria, as well as identification tests. It enables the properties of concrete to be controlled, the regularity of its characteristics to be reinforced and thus its performance and its overall quality in terms of structural durability to be improved.

However, it has been found that the performances of concretes resulting from this approach might not be in line with the targeted objectives of durability and, on the contrary, that compositions excluded by this approach could prove to be interesting.

The usual approach does not enable certain concrete compositions (concretes composed from mineral additions, for example) to be qualified, in spite of being technically interesting and similar in cost. At the same time, evaluations of the potential durability of concretes have progressed and make it possible to provide indicators and criteria (Baroghel-Bouny 2004) that better reflect the performance of concretes than their composition.

The new EN 206-1 standard on the "specification, performance, production and conformity" of concrete therefore incorporates these advances. It puts the durability of the concrete material at the heart of the debate and this approach makes sense for marine structures.

3.5.3.2. *Taking durability into account: the performance approach for concrete*

Structural durability is becoming a major issue with an aging stock of structures, and many studies have been undertaken to optimize structural maintenance and carry out new construction projects (Li *et al.* 2019).

Owners and operators of ports and other marine infrastructures strategically manage the sustainability of structures by taking a proactive approach to maintenance and replacement requirements (Kurth *et al.* 2019). A management plan begins by identifying the infrastructure elements present and to be inventoried, and then the construction materials and protective measures already in place, in order to understand the nature and severity of environmental exposures and their impact on sustainability.

Moreover, a multitude of structural health monitoring options are currently being investigated with a view to ensuring the reliability of concrete infrastructures at various stages of their service life. Therefore, sensors have been developed to monitor parameters such as temperature, moisture, pH, corrosion rate and stress/strain, as well as sensors produced from fiber optic, Bragg grating, piezoelectric, electrochemical, wireless and self-sensing technologies (Taheri 2019).

The performance approach places the durability of the structure at the heart of the designer's concerns. It is made possible thanks to articles 5.2.5.3 and 5.3.3 of the EN 206-1 standard. This new approach sets the durability requirements to be achieved based on, for instance, the aggressiveness of the marine environment. It is a multi-criteria, global, scientifically based and innovative methodology for the specification, evaluation and prediction of the durability of reinforced and prestressed concrete structures (Figure 3.24).

It is based on the concept of sustainability indicators associated with performance criteria and threshold values to be respected. There are therefore no obligations of means, such as in the prescriptive approach described above. Good knowledge of degradation phenomena, allied to the understanding of the various physicochemical mechanisms, has authorized the implementation of this methodology, which is adapted to the requirements of resistance to the aggressions to which the material will be subjected.

The material formulation is considered from the design phase of the project, which makes it possible to work with innovative materials (Xu *et al.* 2019), as well as to formulate materials according to the location and type of the structure (Malyuk 2019).

This approach has been widely used for large civil engineering structures (Millau Viaduct, Vasco de Gama Bridge over the Tagus, etc.). The performance-based approach (Figure 3.24) requires significant changes in the relationships between the construction actors, reinforcement of the responsibilities of each and anticipation of the material problem during the design of the development. It responds to the growing challenge of sustainability for the client; it offers the designer a greater variety of solutions and it allows the company to innovate and enjoy greater freedom in the formulation process.

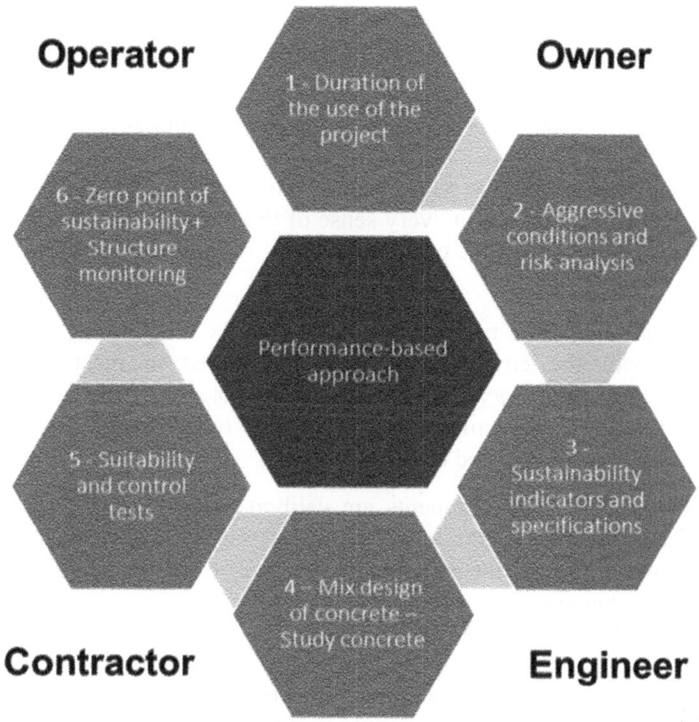

Figure 3.24. *Methodology of the performance approach (from CIMbéton documents). For a color version of this figure, see www.iste.co.uk/ploch/marine.zip*

The performance-based approach thus takes into account the globality of the technical and economical aspects of a structure within its environment and control of the durability of the material. On the other hand, this approach, which deals with the durability of the concrete material, does not take into account the environmental integration of the structures, which is a challenge of sustainable marine environment development .

3.5.4. *The integrative eco-design of the performance approach: perspectives*

The issues are therefore threefold when wanting to eco-design structures interacting with the marine environment:

– the search for appropriate resistance is a fundamental requirement;

– the guarantee of an adequate durability for the envisaged service life is essential. Recent developments in the performance approach aim to better fulfill this objective;

– environmental cofunctions require eco-design of structures.

This combination requires a wide range of knowledge and expertise in order to achieve a sustainable structure in every sense of the word. To be able to achieve all these objectives simultaneously, two fundamental objectives must be integrated in terms of location and duration:

– the material must offer two levels of action for the designer: (i) on the surface, the interface with the environment must be studied to meet environmental objectives; (ii) in the mass, very quickly below the interface, the material must prioritize strength and durability requirements (e.g. the quality of the concrete must guarantee the preservation of the reinforced concrete);

– while sustainability requirements are calibrated for a service life of several decades (usually 50–100 years), colonization of the surface of the structures by living organisms in the marine environment evolves over time. It is paced by the seasons and by changing environmental constants or other environmental effects (Ardizzone *et al.* 1989; Vaz-Pinto *et al.* 2014).

Therefore, it is necessary to accommodate a material that can guarantee a level of performance consistent with the requirements of strength and long-term durability and whose surface characteristics offer a rapidly colonizable support for living organisms.

In Israel, researchers have demonstrated that the pH of concrete can play a role in colonization by living organisms (Perkol-Finkel and Sella 2014). However, they also argue that the texture and macroscopic design of the concrete surface are dominant parameters in the colonization process.

In Norway, the results obtained by *in situ* studies revealed that concrete exposed to seawater undergoes attacks including that of chloride ions. The material and its pH are modified close to the surface from 21 days of immersion. It is also found that the change of surface hydrate phases is independent of the types of binders studied (De Weerdt *et al.* 2014).

These laboratory experiments thus show that the phase changes and therefore the surface pH are “independent” of the nature of the concrete at the core. They also show that the physical parameters (macroscopic relief) are highly significant in the colonization process. Rapidly, the concrete, by interacting with its environment, lowers its surface pH over a time scale compatible with seasonal colonization by living organisms (Hayek *et al.* 2020).

Therefore, the concrete and choice of binder uniquely depend on the strength and durability requirements without contradicting the surface colonization. The latter must, however, take into account a material that is rapidly modified as a result of physical and chemical attacks at the concrete–medium interface.

Finally, authors such as Coombes *et al.* (2013), Harshada *et al.* (2017) and Mos *et al.* (2019) are starting to question the positive or negative effect of biocolonization on the durability of the material based on the location of the marine environment and the nature of the colonization (biofilms, lichens, macroalgae, crustaceans, etc.). It even seems possible that biocolonization by certain species (barnacles, mussels, etc.) may be favorable by minimizing thermal gradients in the material and minimizing the penetration of aggressive ions, such as chlorides and sulfates (Coombes *et al.* 2017).

Mos *et al.* (2019) highlight that concrete fender structures may have a favorable effect on sea urchin (and "calcifying" marine species) settlement by "offsetting" the negative effects of global warming, rising temperatures and seawater acidification. The authors show that concrete performs better than granite or sandstone. The choice of the material is, of course, to be combined with the geometrical complexity of the structure to favor colonization. Ultimately, the artificial concrete structure may participate in the restoration of a species to reduce the negative effects of global warming.

"Biocolonization" is generally perceived as a biodeterioration, but could, in the marine environment, become a "bioprotection" for structures, thanks to the effects of the additional barrier offered by the organisms that have settled on the surface. It is therefore established that the presence of living organisms on the surface of concrete structures in a marine environment will in some way impact the durability of the material.

In the performance approach, the study of the durability of concrete uses predictive models specific to the material and the environment, without taking into account the "living" material–environment interface, which is the subject of eco-design and which adds another dimension. In order to establish a parallel with the methodology proposed for the performance approach, eco-design indicators could complete the general and specific sustainability indicators (Figure 3.25). These eco-design indicators would be linked to the physicochemical characteristics of the surface of the material based on the rate of colonization at time "t" by bacterial, plant or animal organisms (with influence on colonization at time t + dt).

In methodological terms, it follows that a reflection on the eco-design of marine structures must be put in place and formalized in order to define the initial objectives of the project in parallel with the sustainability objectives developed using the performance-based approach. This compatibility and complementarity of the two approaches would make it possible to reconcile the requirement of durability and resistance with a colonization facilitated by the living organisms within a stated aim of maximum ecological gains of the structures. These different aspects treated simultaneously bring about a multi-scale vision of the structure in its globality, as well as of the material which constitutes it. The possible inconsistencies between the different objectives to be reached would then be identified and settled upstream of the project.

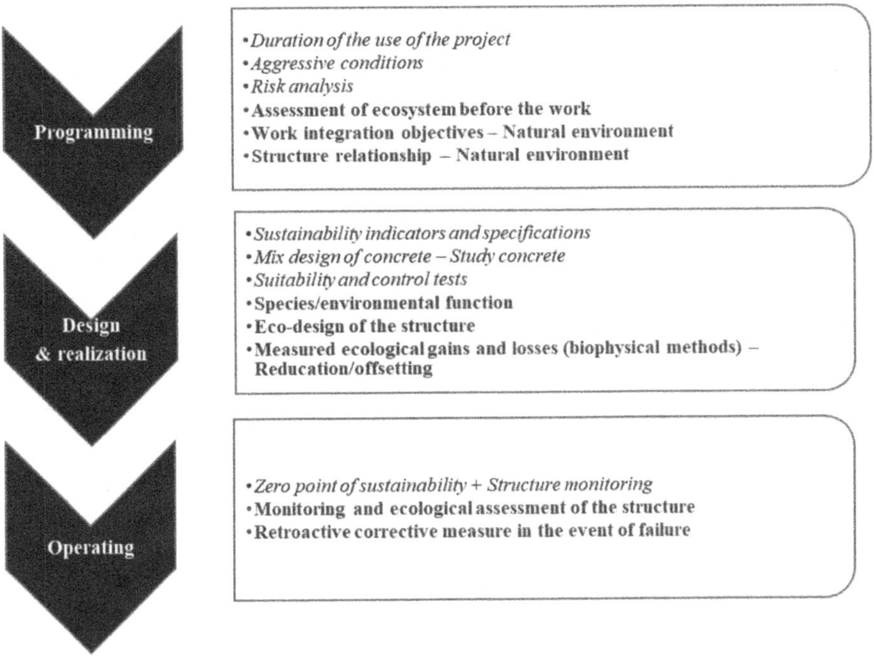

Figure 3.25. *Methodological parallel between the performance approach (in red) and eco-design (in blue). For a color version of this figure, see www.iste.co.uk/pioch/marine.zip*

The need to recycle materials may finally lead to the advent of bioconcretes that will need to meet the technical criteria usually required (technical and functional performance, architectural qualities, durability, safety, ease of maintenance, etc.), as well as environmental or socio-environmental criteria, throughout the life cycle. The future seems open for materials combining bioactivity with durability, in order to create functional and sustainable marine structures, reconciling humans and nature.

4

Evidence Through Experience: Examples of Eco-designed Marine Projects

It is usually presented as a truism that practically any substrate that is immersed in the water is rapidly covered with fauna and flora, the quantity of which would be proportional to the immersed volumes. It is true that at sea, in the same vein as a theorem that Archimedes would surely excuse us for borrowing, "any immersed body undergoes colonization directly proportional to its area".

But what kind of colonization do we want? At the microstructural level, tests on concrete show significant differences in colonization depending on the formula chosen. At the macrostructural level of concrete structures, the question of architectural forms adapted to the needs of the species also arises: which form of habitat (orientation, height, width, depth, color, etc.) for which target species?

In Japan, the work Nakamura (1985) and Kakimoto (2004) summarizes more than a century of scientific studies on the ethological question of the relationship between a species and its (or, rather, their) natural or artificial habitat(s) (the first observations linking the behavior of fish to the shape of a natural or artificial habitat date back to 1903). The objective here is to optimize as much as possible the management intended for "extensive" fishery production using the systemic approach described in previous research work[1].

For our part, we have been interested in this subject since 2008. A first study, in 2008–2009, allowed us to carry out a campaign of immersion of samples in the port of Carnon, in order to compare concretes with biogenic aggregates and standard concretes (Figure 4.1).

1 We refer the reader to the thesis dedicated to this subject by S. Pioch (2008).

a) b)

Figure 4.1. *Photos of colonization of concrete disks after immersion in the port of Carnon: a) with biogenic powder; b) without powder*[2] *(photo: J.-C. Souche). For a color version of this figure, see www.iste.co.uk/pioch/marine.zip*

In subsequent years, from 2013 to 2017, other immersion campaigns were also successively carried out in the port of Grande Motte and, especially, at Ifremer, at the Palavas station in a controlled environment, bringing into play the different parameters influencing colonization (Figure 4.2). This work led to several articles in specialized material or coastal engineering/marine civil engineering publications (Souche *et al.* 2016, 2018a, 2018b, 2019; Hayek *et al.* 2020; Salgues *et al.* 2020).

In parallel, in order to improve the environmental integration of the project to develop the sea, several level 1 demonstrators or developments were carried out by public or private project owners and are presented in this chapter.

First, we will present the eco-design logic of the Mayotte sealine, a pipeline laid at the bottom of the Mayotte lagoon to provide drinking water. Historically, this was the first eco-designed project, carried out between 2007 and 2009.

Since then, completed projects are categorized by major thematic families: port structures, coastal defense structures, artificial reefs, underwater anchoring, etc. In total, we have been involved in nearly 20 projects relating to the eco-design of marine infrastructures.

Laboratory experiments have thus opened up new avenues that could be tested on infrastructures or demonstrators, follow-up of which has enabled their validity to be verified. These projects have, on the whole, enabled knowledge to be obtained as

2 The pink coralline algal colonization is different between samples a and b.

a result of feedback and the results of structural monitoring after immersion. It is by these means that we have enriched our experiences and demonstrated (or sometimes invalidated) the relevance of our initial convictions.

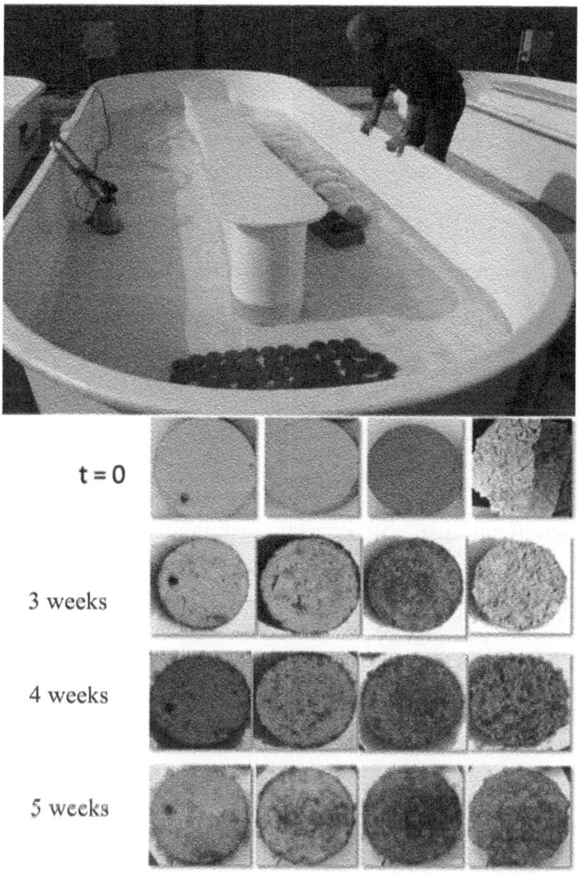

Figure 4.2. *Test tubes in a controlled environment at the Ifremer marine experimental station at Palavas-les-Flots and example results (photo: M. Salgues). For a color version of this figure, see www.iste.co.uk/pioch/marine.zip*

These projects are complemented by other experiences from around the world and by other actors (researchers, contractors, companies) who have been kind enough to share their knowledge and feedback with us for eight developments, for which we are extremely grateful.

For further clarity, the examples cited and the feedback are listed by families of marine structures:

(4.1) Mayotte "sealine" submarine pipeline (2009) (A): the precursor project of a successful eco-design supported by the feedback of Régis Dumay, then Director of Egis Indian Ocean;

(4.2) Bio-inspiration and nature-based solutions for artificial reef design (2009) (B): a new inspiration in structural design;

(4.3) The scope of port eco-design:

(4.3.1) The concept of port eco-design;

(4.3.2) Eco-design of the port of Guilvinec-Lechiagat (C);

(4.3.3) Ecological structures designed as natural grass beds (D); feedback of Fabrice Javel, Director of Port and Marine Development, Safege SAS, eco-design of the port of Antibes (E);

(4.3.4) "Calais Port 2015" port project (F);

(4.4) Eco-design for coastal protection:

(4.4.1) Beneficial use of coastal sediment transport: the Sand Motor project (G);

(4.4.2) The experience of the ECOncrete® company, by Shimrit Perkol-Finkel (H);

(4.4.3) Landscape redevelopment of East Darling Harbour waterfront in Sydney, by Katherine A. Dafforn (I);

(4.4.4) CLI's experience with concrete dike shells: the need for eco-design (J);

(4.4.5) Eco-design of the Brest harbor breakwater, by Anne Rioux, Safege SAS (K);

(4.5) Third-generation reefs and biomimetic anchors, Ajaccio (2016–2018) (L); feedback of Alexandra Agostini, senior assistant at the Corsican Environmental Office, project leader;

(4.6) Eco-design of an artificial island, eco-design of an atypical project (M); vision of the Monegasque government for the offshore extension of the principality of Monaco, by Jean-Luc Nguyen, director of the offshore urbanization mission;

(4.7) Eco-design of organized moorings:

(4.7.1) Eco-design of moorings for recreational vessels in Guadeloupe; a first experience in Deshaies in 2013 (N); eco-design of second-generation biomimetic moorings in Bouillante (2018–2020) (O);

(4.7.2) Eco-design of moorings for large yachts as a solution for the protection of the Mediterranean Posidonia meadow; actions taken by the Maritime Prefecture of the Mediterranean to protect the Posidonia meadow, by André Grosset, deputy unit manager of the *protection et aménagement durable de l'espace marin* (PREMAR, Protection and Sustainable Development of the Marine Area) division, (P); open-sea moorings for large yachts in Corsica: eco-designed anchors (Q) and eco-designed harbor moorings for yachting (R);

(4.8) Eco-design of offshore viaduct piles (S);

(4.9) Eco-design of offshore wind farm project: multi-use perspectives (T).

These references and projects are listed on the global map in Figure 4.3.

4.1. Mayotte submarine pipeline: an initial eco-designed marine structure

This site illustrates the possibilities offered by the eco-design of linear structures such as submarine pipelines, offshore outfalls or submarine cables.

The problem of supplying drinking water to the commune of Dzaoudzi, located on the islet of Petite-Terre in Mayotte, led to the study of a project to restore a submarine drinking water supply (DWS) pipeline with Mamoudzou (Figure 4.4). The length of the 400 mm diameter pipeline was 2.5 km, on a seabed of 0–26 m, with a total cost of 5 million euros. This department, whose lagoon represents 1,400 km² of lagoon with 270 km² of reef area, possesses 2,300 species, including 600 species of fish and 254 species of hard corals (Leone 2014).

The exceptional richness of the seabed obliged the Mayotte Intercommunal Water and Sanitation Union (*Syndicat Intercommunal d'Eau et d'Assainissement de Mayotte*, SIEAM), the project owner, to make the reduction of impacts a strong requirement in its tender documents.

Figure 4.3. *References and feedback presented in this book. Geographical location of eco-designed infrastructures (J.-C. Souche)*

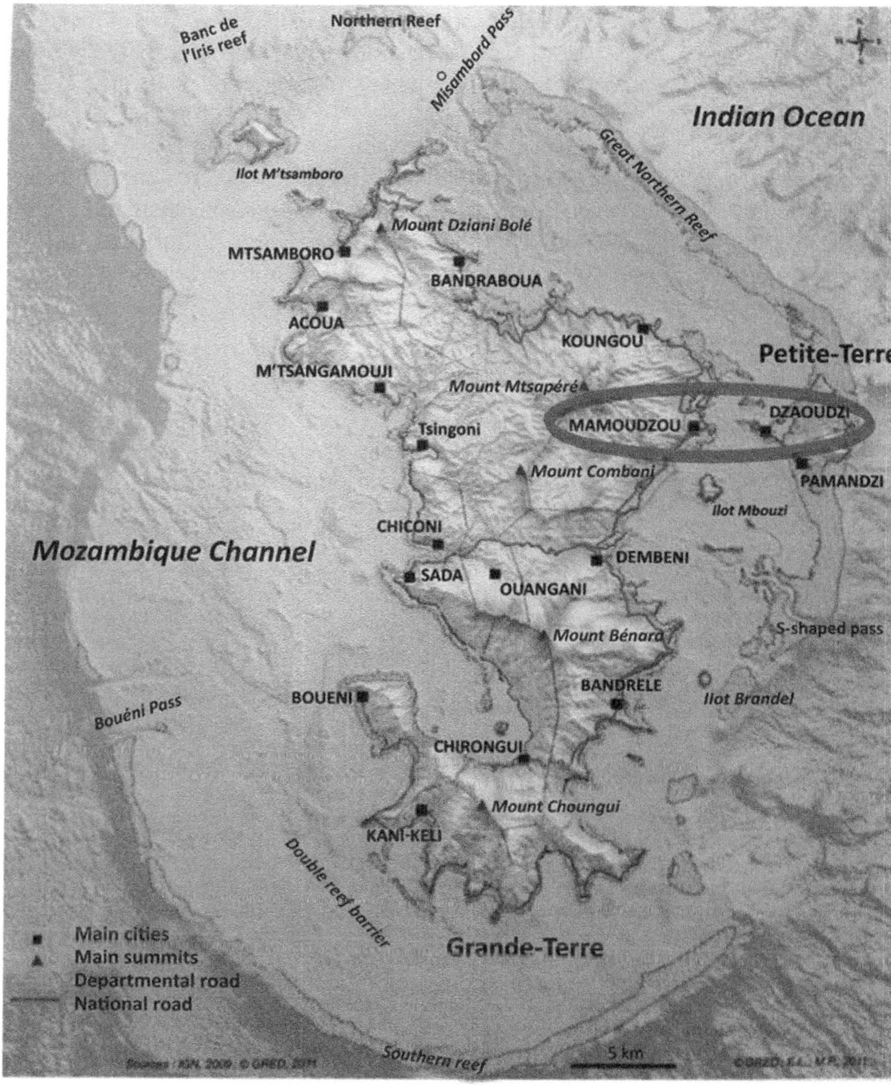

Figure 4.4. *Map of Mayotte and in the blue ellipse the DWS pipeline between Mamoudzou and Dzaoudzi*[3] *(S. Pioch)*

3 Map of Mayotte: major cities, roads and topography (sources: IGN (2009), GRED (2011); design: GRED (2011), Leone (2014)).

Its future classification as a marine park, due to the exceptional size of the closed lagoon (second largest in the world), was also a strong issue, especially for the president of SIEAM, who was a candidate for the presidency of the park being created.

However, as noted by Leone (2014), many local biophysical stresses are anthropogenic, mainly relating to over-sedimentation (terrestrial erosion), mechanical stresses (tourism, fishing), over-fishing of *Acanthaster planci* predators, the prolific coral-destroying starfish and finally global warming. These factors are taking a heavy toll on coral populations (Leone 2014). The avoidance element was also specifically studied, in order not to destroy the coral reefs located along the initially linear project route (Figure 4.5).

Figure 4.5. *Boundary flag indicating the underwater route (yellow dashes) of the structure "zigzagging" between the majority of coral reefs in order to avoid the major impacts of the project (photo: S. Pioch). For a color version of this figure, see www.iste.co.uk/pioch/marine.zip*

In a second step, mitigation measures were considered with the idea of combining the approaches developed during the course of our work on artificial habitat design with the design, or rather the eco-design, of marine engineering structures. For this DWS project, the only parts that could be modified were the concrete weights, from 0.55 to 1.5 t. Their "classic" shape is a concrete cube enclosing the pipe (Figure 4.6).

Figure 4.6. *Cubic concrete weights classically used in marine civil engineering for submerged pipelines (photo: P. Saussol)*

The logic of our approach, which was presented in Chapter 3, allowed us to identify species important for the local food fishery and the developing ecotourism: lobsters and coral fishes, especially serranids. A diving mission gave us a vision of the colonized habitats, at different stages of growth, of these target species. Cross-referencing of this information with local fishermen (underwater, net or line hunters) confirmed the field data on the ethology of the species.

Work on the technical possibilities for the implementation of concrete was also carried out, since were are few resources, quarries or concrete mixing plants on the island. After several tests and models, in particular of flows and the calculation of resistance to tensile forces (the pipe had to remain stable and lie on the seabed), two models were defined and christened "*écocavaliers*" (Figure 4.7; model registered by Egis Eau; inventor, S. Pioch)[4].

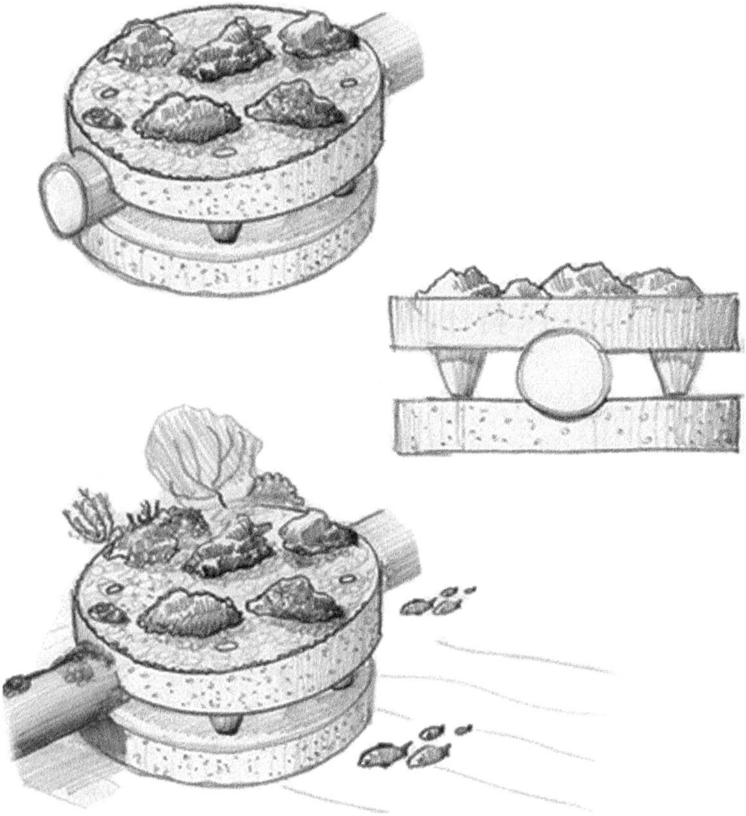

Figure 4.7. *Sketch of an* "écocavalier" *with habitats created to reduce the impacts of the Mayotte DWS pipeline project on the natural coral environment (design: S. Pioch and P. Saussol)*

Based on the depths and the target species present at similar depths and habitats, two shapes were proposed and arranged along the selected least-impact route variant, shown in red and green in Figure 4.8.

4 The Mayotte drinking water supply project, the first eco-designed project carried out between 2007 and 2009, received the first prize in the "Environment and Biodiversity" category at MEDDE in 2010 and the third prize in the engineering category at the 33rd PIANC World Symposium in San Francisco in 2014 (Pioch and Saussol 2014).

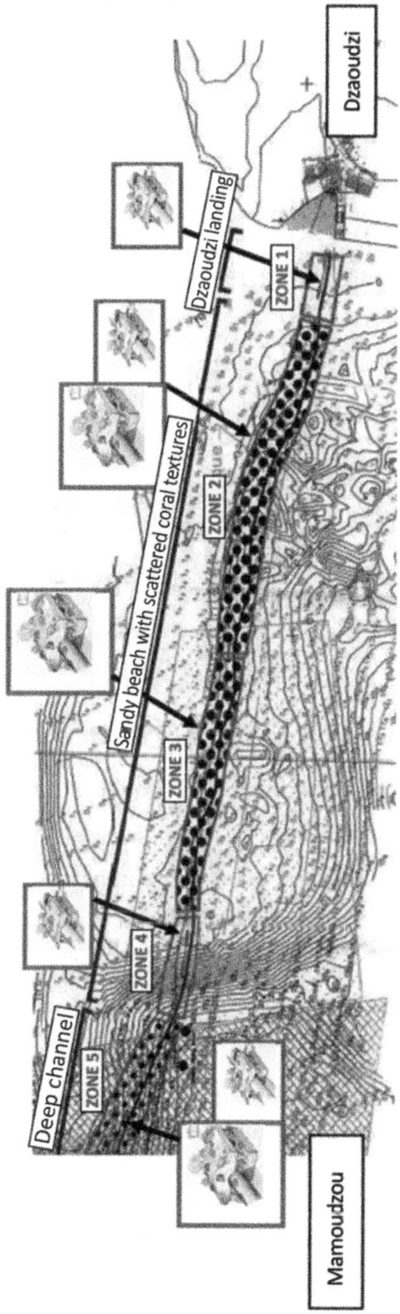

Figure 4.8. *Ground plan of the route and distribution of the different types of eco-designed weights (framed in green or red) according to the ethology of the target species and the natural environment between Mamoudzou and Dzaoudzi (S. Pioch). For a color version of this figure, see www.iste.co.uk/pioch/marine.zip*

After several trials, a mold was developed that enabled the creation of 250 *écocavaliers* arranged along the pipeline (Figure 4.9), with different shapes on the roofs of the structures for different depths and species, as well as surrounding habitats.

During the execution studies, specific provisions had to be determined and developed to settle certain details related to the construction methods, the assembly of the upper and lower prefabricated parts, the cohabitation on the roof of the specific devices for the reception of biodiversity and the passage of the chains of maintenance of the *écocavaliers*. Finally, a specific reinforcement of the *écocavaliers* was designed to guarantee the resumption of the forces applied to the pipe and the *écocavaliers*.

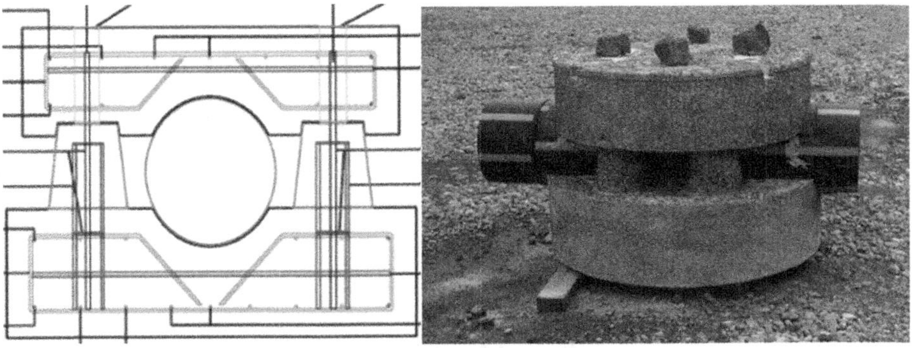

Figure 4.9. *Reinforcement of the pipe weights, functional drawing (design: J.-C. Souche) and the first* écocavalier *pilot prototype (photo: S. Pioch). For a color version of this figure, see www.iste.co.uk/pioch/marine.zip*

Ecological monitoring was stopped after six months due to lack of funds. It revealed about 40 species on the écocavaliers, whereas only half of that number were counted on the old pipeline, which had been in place for more than 30 years (Figure 4.10). It is also noteworthy that the environmental studies were approved by the authorities, without any major comments, and, in fact, with encouragement for this type of innovation for the design of future projects.

Since then, a variation of this system has been installed in the Mediterranean for the Propriano outfall in Corsica (2011) and another is the subject of a project in progress for the Monaco outfall in 2021.

Figure 4.10. *Monitoring of the colonization of structures in 2009 after one month of immersion (photos: S. Pioch)*

"Between 2005 and 2010, the concept of artificial reefs began to emerge in the Indian Ocean; the main problem was their financing. I was convinced that we, Egis and more generally the engineering companies, had to play a major role in preserving marine biodiversity and protecting the marine environment. As a designer of marine structures, we were in a privileged position to propose development solutions, while reducing the impacts on the sea. It was, in fact, the combination of those two elements, the financial constraints and the environmental conviction, that gave rise to the idea of eco-design.

Indeed, the question of financing meant that we had to be pragmatic. At Reunion Island and Mayotte, the marine works that were commissioned by Egis at that time were mainly offshore pipelines. They were either effluent discharge outfalls from wastewater treatment plants or drinking water transport pipelines. We therefore decided to transform these structures, which were having a negative impact on the environment, into benefits for the environment. The great advantage of working on structures whose primary vocation was not, strictly speaking, 'environmental' was that we quickly realized that the cost of eco-design was marginal in the overall budget of the project owner. It was, therefore, a winning strategy for biodiversity and one that could be valued by our clients.

The most emblematic project of the time was the Mayotte 'sealine', the precursor of this approach. Sylvain Pioch and Jean-Claude Souche, who were at Egis, actively participated in this project. The problem of supplying drinking water to the commune of Dzaoudzi, located on the island of Petite-Terre in Mayotte, led to the construction of an underwater drinking water supply pipeline with Mamoudzou. The project owner was SIEAM, and its exceptional wealth prompted it to make the reduction of impacts a strong requirement of this operation.

A total of 250 *écocavaliers* were made and placed along the pipeline, with different shapes according to the target species (Figure 4.11). After six months, ecology monitoring revealed the complete success of the marine colonization, with more than 40 species already present on the *écocavaliers*, for a marginal additional cost of 1% of the total project budget. This result was only possible because eco-design was initiated from the outset of the project and because each actor played their role fully at each stage.

Although, from 2000 to 2010, I was particularly sensitive to the protection of marine fauna, the first eco-designed projects convinced me and all the Egis teams to develop the concept and take it further.

Figure 4.11. Ecocavaliers *for ballasting the Mayotte marine pipeline (photo: R. Dumay)*

Thus, Seaboost was founded in 2013. This Egis subsidiary specializes in the design, manufacture, installation and scientific monitoring of implemented technical solutions. The developments are numerous and range from artificial reefs to eco-designed technical structures (e.g. weights for pipelines or yachting moorings), including cladding modules for submerged structures (notably port or offshore eco-design). Based on the concept of biomimicry, these customized solutions hinge on the reconstitution of specific natural or artificial habitats, with the aim of stimulating key ecological functions for the balance of the environment and/or the support of uses such as fishing or tourism. R&D and innovation are behind most of the solutions proposed, such as the next-generation artificial reefs in Qatar, the 3D modules installed in the Calanques National Park to restore degraded ecosystems, and the modules for covering port structures such as quays and dikes in the port of Marseille."

Figure 4.12. *Immersion of artificial reef by Seaboost (photo: R. Dumay)*

Box 4.1. *An initial operation in which Régis Dumay*[5]*, then Director of Egis Indian Ocean, took part*

5 Régis Dumay is currently Deputy Managing Director of Egis.

4.2. Bio-inspiration and nature-based solutions for artificial reef design

According to the Food and Agriculture Organization (FAO) the definition of an artificial reef is limited to "any submerged human construction, integrated into the ecosystem, whose design objective is to protect and develop aquatic fauna and flora". However, the aesthetic question of an artificial object placed within a still "natural" environment, the seabed, seems to be just as important.

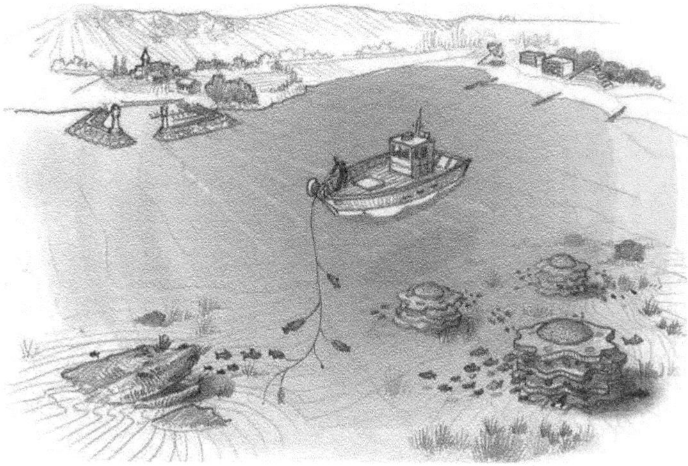

Figure 4.13. *Developing rocky seabeds with artificial fishery habitats ecologically and aesthetically integrated into the underwater landscape (concept: S. Pioch, design: J.-L. Feron)*

In fact, meeting the "classic" needs of development responds to precise rules of art, currently the subject of specifications, that never integrating underwater landscape aesthetic values; and when projects do have the financial means to fully assume this ambition, the morphologies of submerged structures such as reefs or artificial habitats are resolutely human and terrestrial in tone: cubes, trapezoids, beams, cylindrical nozzles. They are very far from the natural soft shapes usually encountered under the sea.

Shipwrecks are perhaps a little better integrated, due to their imagery linked to a "sea" culture, but do not constitute an ecological summit when considering landscape or functional issues. It would therefore seem that a landscape approach to artificial habitats, for example, through bio-inspiration or mimicry with natural reefs, would enable a better visual integration of these infrastructures under the sea and, undoubtedly, a better acceptance of the projects by stakeholders and the general public (Figure 4.14). At the same time, these biomimetic structures are now proving to be more ecologically efficient in many respects. So, if we can create something beautiful in order to do something effective and better, why deprive ourselves of it?

Figure 4.14. *Example of an artificial habitat mimicking a natural drop-off sheltering a Mediterranean red coral cave through "ecomimicry"; on the left, a natural rocky coral drop-off; on the right, an artificial habitat attempting to approach it (design: S. Pioch, drawing: J.-L. Feron). For a color version of this figure, see www.iste.co.uk/pioch/marine.zip*

4.2.1. *The value of creating artificial habitats*

The most often sought objective in the establishment of such structures is to favor fish species in order to "create new good fishing spots" or to ensure vital functions for their development.

The purpose of these artificial fisheries habitats is to: (i) facilitate the settlement of species for fishing; (ii) restore the seabed; (iii) increase fishery resources; (iv) maintain and develop biodiversity by creating diverse habitats; and (v) create biological corridors.

Some of these structures present interesting results. For example, in the Côte Bleue marine park (Carry-le-Rouet), on a soft Mediterranean seabed, before the installation of reefs, production of biomass was 6 kg/year/m^2; six years later, this production was 25 times higher and reached 150 kg/year/m^2 (Figure 4.15).

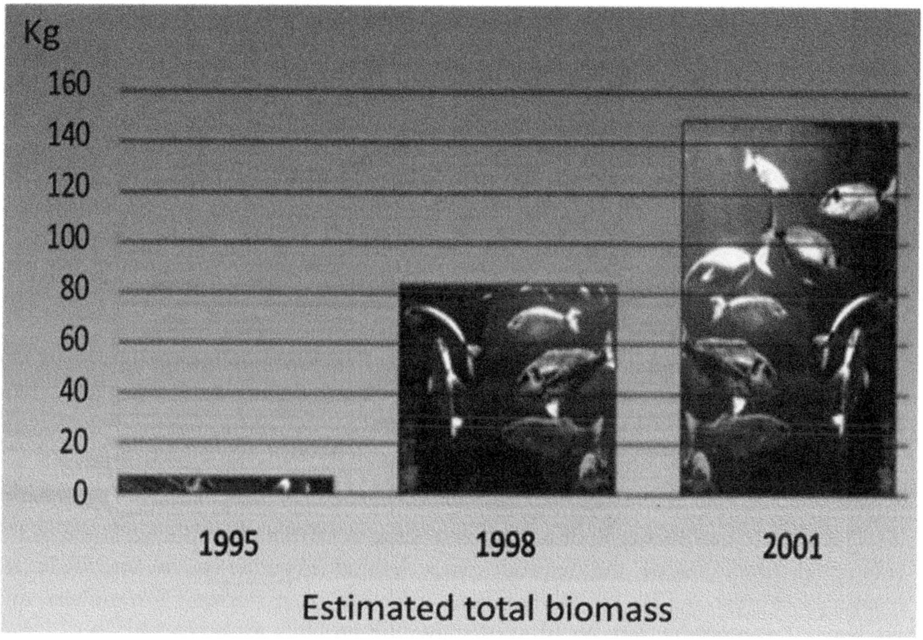

Figure 4.15. *Example of increase in total biomass through the installation of artificial reefs, Côte Bleue marine protected area, Bouches-du-Rhône (Charbonnel* et al. *2002)*

Of course, it is rather surprising from an ecological point of view to compare the production of a soft substrate with that of a newly created hard substrate (creation, not ecological restoration, in reference to the definitions previously stated), but, depending on the initial objectives, it may be pertinent.

4.2.2. *The issue of fish habitat creation*

First, this issue is about mimicking nature in order to optimize it. Mimicking natural seabeds that have been degraded or are absent facilitates the settlement of target species for fishing; this is a nature-based solution approach. Specific habitats must also be recreated to provide suitable shelter for species (Figure 4.16). The concept of ecomimicry is fundamental to returning fish to the habitats and functions they are accustomed to in their natural environment.

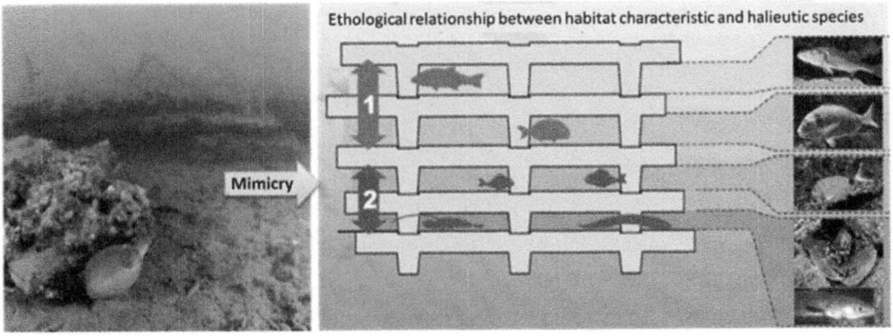

Figure 4.16. *Schematic of the design of artificial habitats inspired by nature through biomimicry, where each stage is designed according to the needs of a fish species in a habitat (right), inspired by the natural habitats frequented by these species (left), here a biogenic rock (S. Pioch)*

The objective of this type of habitat is to seek to artificially replicate some of the ecological functions of the natural rocky habitat in order to restore fisheries resources (Figure 4.17): (i) nourishment: creation of a nursery for grazers and predators; (ii) reproduction: creation of spawning grounds for reproduction of target species; (iii) protection: creation of constant or temporary shelters for sedentary or migratory species.

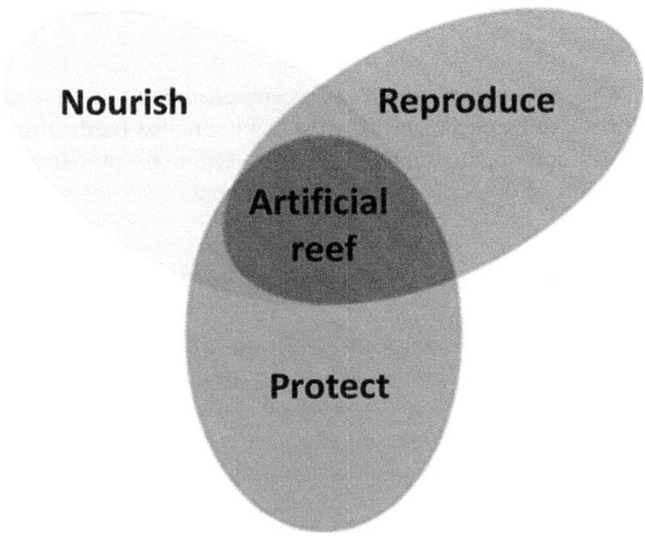

Figure 4.17. *Ecological functions for the maintenance and development of target species of an artificial reef (S. Pioch)*

We have already detailed the design process, but we here provide a reminder that it is broken down into several steps, in order to establish a relationship between the natural habitats observed in the implementation area, the artificial habitat and the local target species:

– identification of the biophysical and socio-economic characteristics of the environment;

– determination of the local ethology of the target species (scientific works, dives and surveys);

– co-construction (fishermen, divers, etc.) and elaboration of a sketch, taking into account local knowledge;

– validation of the project's ecological and social objectives and establishment of a final manufacturing guide.

Two first prototypes of biomimetic "ecoreefs", that is, combining shapes adapted to the needs of target species, a rough shellfish concrete[6] and an aesthetic inspired by natural forms (Figure 4.18), were immersed off the coast of Agde, a commune in the Hérault department, in 2009. This *in situ* application of artificial reef eco-design was carried out within the framework of a public project management contract with the municipality of Agde for the immersion of 99 first-generation artificial reefs: concrete nozzles (DN 1000/1600), steel cubes, decommissioned EDF poles and manholes for concrete stormwater networks.

These two experimental prototypes[7] were made of shellfish concrete (recycled shellfish waste supplied by the company Coved). Their original design was based on the study of the ethological needs of local target species such as: wolffish, sars, conger eels, forkbeards, slippers and lobsters. The shapes were inspired by natural rocky landscapes (local shelter), in application of the methodology presented in Chapter 3. They were financed in partnership with the General Council of Hérault, and were christened "second generation" with innovation features such as: (i) resumption of efforts toward shellfish concrete structures, (ii) the search for an artificial habitat adapted to the functional needs of the target species and (iii) landscape integration by mimicry of the shapes of natural habitats.

After seven years of immersion, in 2016, the nozzles installed in the seabeds similar to the two prototypes (-13 m) showed strong signs of scouring (peristructure depression and strong siltation, up to disappearance under the sediment for some), while the two ecoreefs remained stable. The hydrodynamic flows were calculated to

6 Development of concrete formulation by Jean-Claude Souche, IMT Mines Alès.

7 Manufactured by the company BEC Frères and immersed by the company BUESA.

avoid local scouring or sinking effects (see Chapter 2 for a description of the necessary marine civil engineering studies).

a) b)

Figure 4.18. *a) Immersion of two prototype structures off Agde in September 2009; b) control of the immersed structure by S. Pioch by snorkeling (photos: S. Pioch (left) and M. Foulquié (right))*

The target species, such as common white seabream (*Diplodus sargus*), conger eel (*Conger conger*), gilthead bream (*Sparus aurata*) and bass (*Dicentrarchus labrax*), whose ethology guided the shape of each stage, are regularly caught by the small-scale fisheries, who have been proven to be exploiting these artificial reefs (Figure 4.19).

Monitoring of the colonization was carried out by the Adena organization over a period of two years. The conclusions show that the target fish species were present and in greater quantities than were found on conventional artificial reefs. However, the colonization by fixed species (worms, mussels, etc.), although initially rapid on the shellfish concrete following immersion, reached a quantity and diversity plateau similar to that observed for the "standard" concretes used for the nearby immersed nozzles. Shellfish concrete did not bring any added value in this respect. Thanks to the optimization of manufacturing costs (as a result of this pilot project), the ecoreefs do not present any significant additional costs, compared with prefabricated concrete elements of comparable volumes that are not designed for target species.

Since this project, variants of these ecoreefs have been immersed off the coast of the commune of Propriano in Corsica, in 2011, as well as in the Cherbourg harbor in the English Channel (2015) and finally in the framework of the RexCor program for the restoration of the seabed in front of the Marseille wastewater outfall, in 2017, by the company Seaboost, which is operating the registered model (see F. Martarèche's personal vision in Chapter 3, Box 3.5). Finally, 200 ecoreef modules were immersed as a compensatory measure for the negative impacts of the New Coastal Road (*nouvelle route du Littoral*, NRL) at Reunion Island in 2017.

Figure 4.19. *Adult seabream (*Diplodus sargus*) during the breeding season, settled in a stage of the ecoreef that serves as a shelter and spawning ground (photo: S. Pioch)*

4.3. The scope of port eco-design

Following these first experiments in the "open sea", development of the eco-design concept has continued in the context of port structures where the major problem of water quality arises.

Indeed, there would be no point in equipping ports where pollution from antifouling paints[8] or the untreated run-off from peripheral anthropogenic activities (wastewater, run-off from careening areas or parking lots, etc.) would continue. Worse, this would reduce the vital potential of the juveniles often present in these areas of calm environments, similar to lagoons, which constitute artificial nurseries.

We will not detail all the studies resulting in the definition of the development options proposed to illustrate the overall concept described in Chapter 3, but rather report below on some representative studies that have been conducted.

4.3.1. *The concept of port eco-design*

The concept of port eco-design, which is close to a systemic vision, targets the consideration of this artificial environment as an ecosystem, where species will colonize structures according to their ethology. Thus, the parts close to the open sea will be used for adult species, the intermediate parts (outer harbor, first pontoons) for young adults and juveniles and the seabed of the harbor areas for juveniles (Figure 4.20), subject to good water quality, of course.

8 Paint containing biocides designed to prevent aquatic organisms from attaching to the hulls of vessels or other submerged objects.

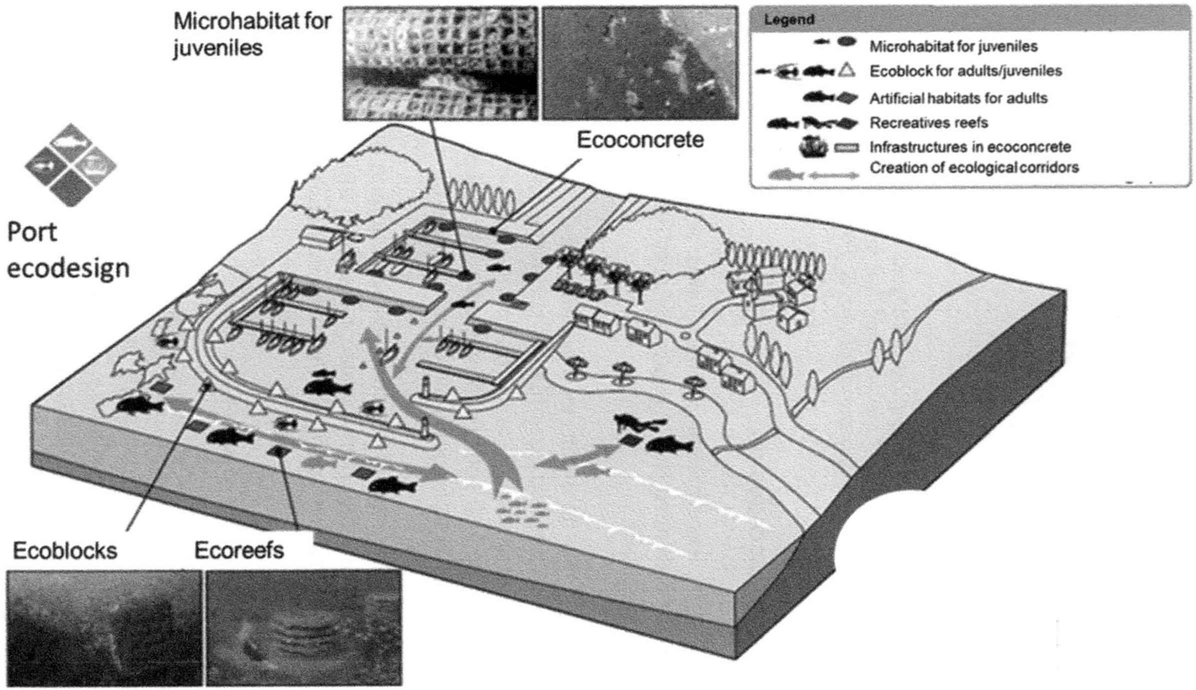

Figure 4.20. *Proposed overall concept of port eco-design from 2009 where the port is a new ecosystem which can enhance the artificial habitats created from the ethology of the colonizing species (design: S. Pioch, drawing: J.-L. Feron, Egis Eau). For a color version of this figure, see www.iste.co.uk/pioch/marine.zip*

4.3.2. *Eco-design of the port of Guilvinec-Lechiagat*

In order to reduce the impacts of the works, within the framework of rehabilitation studies or port extension works, it is important to carry out a diving inventory of the port species as well as a study of their behavior: position in the water column, hunting or resting activity, escape or sheltering behavior, etc. The "port ethology" map proposed below was made by snorkeling in the port of Guilvinec-Lechiagat, and illustrates this approach (Figure 4.21). Note the presence of different stages of development according to a particular gradient: adults towards the entrance, juveniles in the sheltered and shallow, calm parts, and intermediate stages in between.

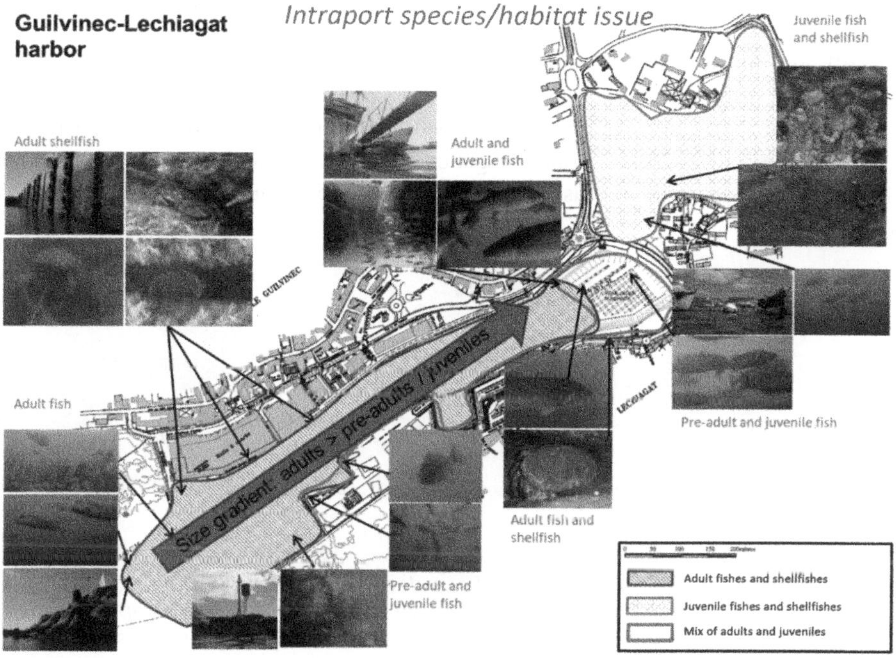

Figure 4.21. *Map cross-referencing ethological information and the different artificial port environments sheltering fishing target species: sea bass (adults and juveniles depending on the area), edible crab, shrimp, red mullet and pollack (pre-adults and juveniles) (graphic design and underwater photos: S. Pioch, background map: commune of Guilvinec (2010)). For a color version of this figure, see www.iste.co.uk/pioch/marine.zip*

The harbor is colonized differentially according to the substrates and habitats, often unintentionally created by humans, forming a shelter for some species by “happy coincidence”.

Moreover, these artificial shelters are extremely valuable in helping to design, this time intentionally, functional habitats during the design of eco-designed structures.

Biomimicry (nature-based solutions) with peripheral natural habitats will complement architectural considerations for eco-design.

4.3.3. *Ecological structures designed as natural grass beds*

Still on the subject of port achievements, the photos below are from the experimental program carried out in the port of Fos (Bouches-du-Rhône), from 2011 to the end of 2015, as part of the Girel eco-restoration program (financed by the Rhône-Méditerranée-Corse Water Agency and the Grand Port Maritime de Marseille). The operational feasibility of eco-design, even at the experimental stage, is confirmed.

a) b)

Figure 4.22. *Pilot applications of eco-designed dock for port structures with the “sea urchin” (a) and “seagrass” (b) types; the bio-inspired approach can be appreciated in these photos (design: S. Pioch, photos: Egis Eau)*

The company Suez Environnement has, in parallel, developed works for the ecological restoration of algae of the Cystoseira genus, an engineer species of the small Mediterranean coastal seabeds. Initial tests having been conclusive and the eco-design of engineering structures integrating specifications relating to roughness, layout during installation and seeding, was implemented in 2016 (see Box 4.2) and then within the framework of the Port Vauban (Antibes) works in 2017.

"The algae of the genus *Cystoseira* play a major ecological role in the functioning of rocky ecosystems in the Mediterranean. Qualified as engineer species, they form real forests, places of refuge and feeding for many species. Within the *Cystoseira* genus, *C. amentacea* is still abundant on the rocky Mediterranean coasts, but is subject to the pressures of urban waste and coastal development. Within a general framework of solutions development for ecological enhancement of marine infrastructures, the objective of the project is to test the validity of technical specifications applicable to the construction of dikes designed to increase the potential of Cystoseira reception. Benefiting from positive feedback on the restoration of Cystoseira, Suez Consulting and the Mediterranean Institute of Oceanography proposed to adopt a two-step approach aiming at obtaining a colonization of the dike by Cystoseira.

The first step consisted of maximizing the distance that could be colonized by Cystoseira by selecting boulders with significant roughness and arranging them in such a way as to obtain slight surface slopes. These very simple arrangements did not incur significant additional costs on the scale of the operation (only a few percent in the context of the work carried out by the company TP Spada in 2016). In a second phase, Suez Consulting, with the support of the GIS Posidonie, the port team and the city of Antibes, carried out a seeding of the dike with Cystoseira, the Cystore® process having already provided encouraging results in the Calanques National Park and for the dykes of the Grand Port Maritime in Marseille. This seeding took place during the summer of 2017, covering a distance of about 300 m. The monitoring of Cystoseira recruits (growth and survival) began in the fall of 2017 and continued until the start of summer 2019. Despite good survival rates, recruits remained very short, due to very high grazing pressure by herbivorous fish (salema). Only the blocks offering a strong macro-roughness, limiting the accessibility to herbivores, allowed the maintenance of Cystoseira. The coherence of the approach and the strong involvement of the stakeholders, for all their diversity, were real sources of satisfaction that encouraged us to continue the efforts for a species of great ecological value. We still need to better understand and control the parameters favorable to the settlement of the species, but one of the prospects that is currently emerging is the design of prefabricated blocks specially adapted to their needs."

Box 4.2. *Feedback from the Suez Environnement company on the eco-design of structures adapted to Cystoseira algae species by Fabrice Javel*[9]

4.3.4. *"Calais Port 2015" port project*

The following is an example of an eco-designed port structure study for a large marine port, part of the "Calais Port 2015" extension project. This project has been

9 Director of port and marine environment development, Suez Environnement. The Port Vauban project was carried out with the financial assistance of the Rhône-Méditerranée-Corse water agency.

the subject of sketch studies since 2003 and includes a large dike reclaimed from the sea. These works must meet the needs of future generations of ferries and ro-ro ships and cope with the increase in traffic expected by 2020–2025 in the Channel.

Figure 4.23. *Photomontage of the Calais Port 2015 extension project*[10]

Construction of the works began in early 2016. They consisted of a dike of about 3,000 m in length, built to shelter a 110 ha body of water. The pass was dredged to the coast, –13 m NGF. A portion of 90 ha of new land was reclaimed from the sea, thanks to the fill from the dredging of the basin. A rail terminal on the quayside was installed between the two basins of the port. The overall cost of the Calais Port 2015 project for the company Bouygues TP, which carried out the work, was 652 million euro (Figure 4.23).

This sandy zone is home to species specific to the Channel and North Sea. Additionally, the powerful tidal ranges and the corollary currents required that different "in-water" structures be considered in order to maximize the ecological benefits and reduce the impacts of the project's right-of-way. There was no question of transposing the tools and solutions developed in the above-mentioned projects.

The species are mainly flatfish (the largest sole (*Solea solea*) nursery in the Channel) and sea bass (*Dicentrarchus labrax*), which are abundant in the zone due to the large gravel pits and shipwrecks that are favorable to gregarious behavior during spawning, as well as large crustaceans (lobsters, edible crabs).

Floating habitat systems for fauna, fixed by flexible anchors on the dike, were proposed to follow the movements of the water line under the effect of the tides. They were completed using innovations, developed in Japan, in the form of dike blocks integrating flat fish supports, serving as nurseries and reinforcing, at the same time, the stability of the blocks as part of a "win-win" approach.

10 Photo: CCI de Calais, https://www.meretmarine.com/fr/content/calais-port-2015-le-conseil-regional-vote-la-poursuite-du-projet.

The ethological (stage, species, vital needs) effects, linked to hydrodynamic conditions (surface agitation, currents, sediment reworking, etc.), were studied in particular and are presented in Figure 4.24. This enables us to understand the spatial logic of eco-design and the types of structures that were recommended according to the types of needs of the target colonizations (agitation conditions, depth, surrounding substrate, type of colonized shelter, etc.).

Three sections have been designed according to the type of target species, the development stage and the expected ecological function(s) (depending on the environmental conditions, depth, current, sediments, etc.):

– the first section of the dike (in blue in Figure 4.24) is dedicated to a flatfish nursery (sole, plaice, etc.), in place naturally before the start of the project and observed *in situ* by diving, with the design of dike blocks arranged to mimic the natural habitats favorable to this priority function;

– the second section (orange) is a zone of mixed development stages and benthic species (mainly mullets, sea bass, rays, lobsters, spiders and conger eels) with dike blocks and artificial habitats adapted by their location and shapes;

– the third section (green) is dedicated to two adult species, sea bass and lobster, for their reproduction (sea bass spawning ground) or the maturation of their eggs (lobster), with aerated dike blocks or provisions favoring the spawning function.

Agitation conditions (tide, swell and current) and the presence of both spawning adults of sea bass and lobsters, as well as a large flatfish nursery, typical and unique nationally, influenced this triple segmentation.

It can also be understood here that a fixed approach (a type of design "plastered" along the entire length of the dike, without taking into account tides or local ecological typologies), or the addition of "post-design" modules to the dike blocks, would diminish the effectiveness (and especially the sustainability of the proposed ecological solutions), at the risk of green washing, the perversity of which we saw in Chapter 1.

Effectively, this would necessitate a double compromise, requiring the running of a project to artificialize the seabeds and, in the short term, ecological solutions "in front" to justify it.

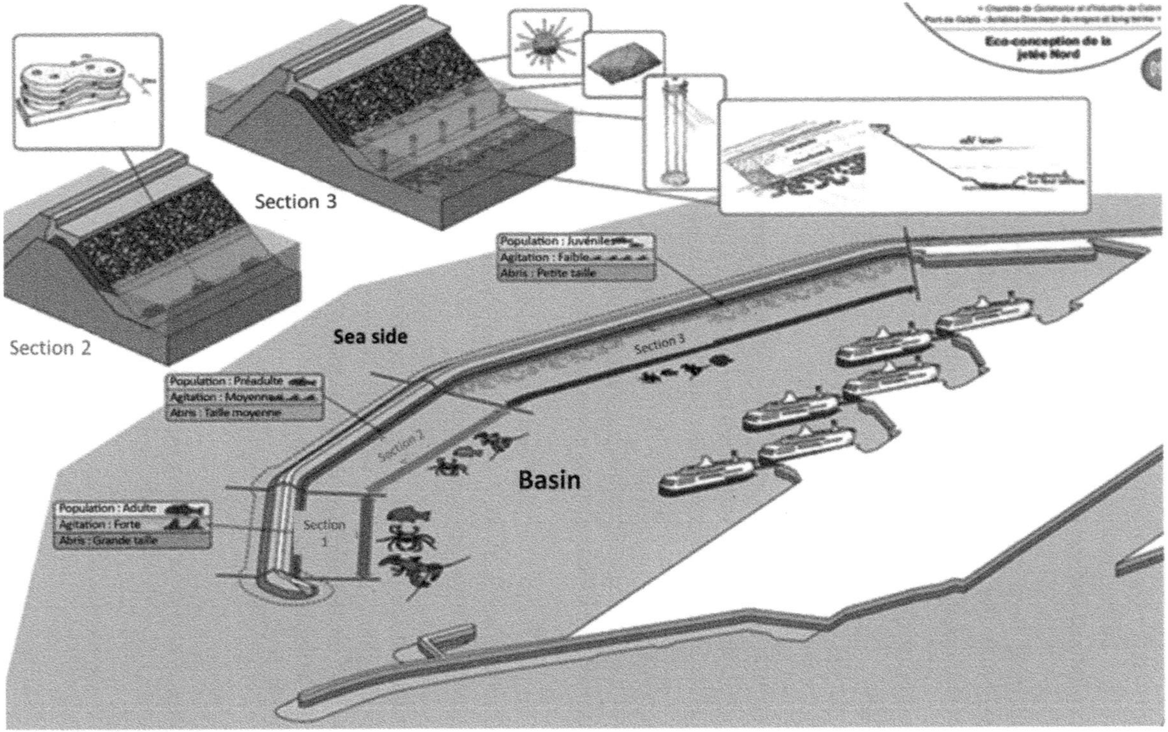

Figure 4.24. *Three-part segmentation of the north jetty for the protection of the Calais Port 2015 zones, according to the types of target species, their development stages and their habitat requirements, based on the natural habitats present in the Eastern Channel (design: S. Pioch, drawing: J. L. Feron, Egis Eau). For a color version of this figure, see www.iste.co.uk/pioch/marine.zip*

Figure 4.25. *Photo of seabass shelter, in a typical habitat adapted to resting and protection needs (photo: S. Pioch)*

4.4. Eco-design for coastal protection

The coupling of a strong human presence on coastlines with the changing climate, and the consequent coastal erosion, has resulted in a need for port (dikes) and coastal (swell attenuators) protection infrastructures.

This problem has given rise to various research projects around the world.

In Europe, the EU's THESEUS project dealing with the design of greener dams has demonstrated that environmental provisions integrated at an early stage of the design process allow for greater freedom of innovation. The positive effect on the environment improves the performance of the structures (bioprotection), at an acceptable cost, as discussed in section 3.1 of this book (Firth *et al.* 2014).

In the Netherlands, the concept of working with nature has been applied to various projects, in particular relating to the Sand Motor intervention. This project demonstrates that the proper use of coastal sediment transport enables erosion to be avoided, provides better protection and has increased the environmental richness of almost 5 km of sandy coastline. Natural ecosystem services which benefit humans (fisheries, carbon storage, ecotourism, etc.), have also been accounted for, helping to balance the costs of coastal erosion protection.

In Israel, the experience of the ECOncrete® company demonstrates that the extension of environmental functions in the design of marine structures is a growing concern in development projects. There is, therefore, a need to develop research based on pilot demonstrators acting as real-life tests to evaluate the ecological and economic cost/benefit. Indeed, the body of knowledge is still weak in providing criteria to help decision-makers, engineers and companies make choices (Sella and Perkol-Finkel 2015). While it is possible to modify construction materials or surface conditions to create ecosystem-friendly diversity, these modifications should not negatively impact the technical functionality of the structures (Perkol-Finkel *et al.* 2018). Moreover, the ecological gain must be systematically verified through monitoring to validate the success of the initial objectives: locally important target species, specific ecological functions, manufacturing costs, no maintenance of the self-managed system, etc.

In Australia, the development and redevelopment of the East Darling Harbour area (Sydney) and its waterfront highlight the aesthetic quality of the landscaping used. They were based on the use of blocks of sandstone, a natural local rock, which were also arranged in such a way as to mix the size and volume of the cavities of the protective seawall. These blocks, which replace the former vertical concrete dock walls, promote the intertidal habitat of the dike, while providing an aesthetically pleasing structure and recreational space for the people of Sydney (see Figure 4.29).

In France, the French company Concrete Layer Innovations (CLI), part of the Artelia Group, a major world player in concrete shells for embankment dams, promotes successful environmental solutions. Their ECOPODE blocks aim for better aesthetic integration for the above-ground parts of dikes, as well as in terms of roughness for the submerged parts. These solutions are providing new information on the comparative effects on colonization, with structures made of conventional concrete or even rockfill blocks.

4.4.1. *Beneficial use of coastal sediment transport: the Sand Motor project*[11]

There is a growing scientific and engineering interest in exploring how natural processes can help to provide management solutions to resolve the degradation and/or vulnerability of marine, coastal and estuarine environments. While climate change and the associated sea level rise are often suggested as major risk factors, anthropogenic drivers, such as marine resource exploitation, subsidence, reduced sediment supply, reclamation and coastal squeeze, are often more dominant risk factors. Using natural processes to deal with these risk factors presents a non-trivial challenge.

11 Section written by Professor Emeritus Marcel Stive, Department of Hydraulic Engineering, Faculty of Civil Engineering and Geosciences, Delft University of Technology, the Netherlands.

Nature-based solutions require a comprehensive approach to resolve the science and engineering aspects. These solutions address societal needs for flood, draught and environmental protection and restoration, sustainable infrastructure and food production while providing additional benefits in terms of increased biodiversity, CO_2 sequestration and recreational and economic values. Furthermore, nature-based solutions are increasingly designed to be better adaptable to changing circumstances, to be more cost-effective and more impactful than our traditional solutions.

The Sand Motor on the Netherlands coast is such a nature-based solution par excellence. It is designed to provide flood protection over the next 20 years in combination with the creation of recreational and nature values. This project required deep knowledge of the physical and biological processes and innovative governance, plus institutional, business and finance models. It also required an inclusive collaboration between various disciplines and sectors, and the engagement of all relevant stakeholders from the initiation phase of the project all the way towards final implementation and construction (Figure 4.26)

Figure 4.26. *General view of the Sand Motor project site (photo: M. Stive)*

The Sand Motor now inspires a world in which nature with its capacity to be destructive is actually our best friend in building sustainable and resilient coasts and safeguarding our coastal towns and economies while improving the environment and the ecology and mitigating climate change. It is time to learn, replicate and scale up these nature-based solutions globally. We must start to educate a new generation of engineers, designers, ecologists, policy makers, politicians and managers on the principles of nature-based solutions. We must develop credible narratives and the business cases and examples of successful projects, implement larger scale projects worldwide and evaluate these and disseminate the experience and knowledge gained. We must involve communities in the planning and implementation and equip them with knowledge and financial means in order to scale up and sustain such solutions, working side by side with governments, universities and the private sector.

4.4.2. *The experience of the ECOncrete company*[12]

The construction and expansion of ports and marinas significantly increases the amount of artificial substrate in the marine environment. Such structures offer new substrate for colonizing organisms, but too often, they do not support similar species assemblages as those of natural coastal and marine habitats. Traditionally, port breakwaters are composed of standard concrete or rock armor that is typically formed of smooth/featureless units, which offer limited habitat value. Such artificial structures usually promote low-diversity species assemblages and are dominated by exotic and nuisance species. This is mainly due to substrate and design features including harsh surface chemistry, steep slopes, low structural complexity and high homogeneity, all of which are rarely found in natural habitats.

In 2012, a new port was constructed at the Haifa Harbor (Israel). The breakwater was made of cube-shaped portland cement armor units called “Antifers”. In parallel, ECOncrete® deployed an environmentally sensitive alternative, of bio-enhanced Antifers (ECO Armor Blocks), designed to enhance biodiversity and productivity while complying with all of the constructive and engineering requirements of marine construction. The environmentally sensitive design differs from standard Antifers on three levels – concrete composition, surface complexity and macro-design.

These three elements work in synergy to make the bio-enhanced breakwater constructed from ECO Armor Blocks a preferable habitat for marine organisms, compared to standard Antifers, as they mimic features and biological niches of natural marine habitats, providing improved ecosystem services and supporting a more diverse and rich species assemblage, thus reducing the ecological footprint of the structure (Figure 4.28).

12 Section written by Dr. Shimrit Perkol-Finkel, CEO and Scientific Director of ECOncrete Tech LTD, Israel and USA.

The ECO Armor Blocks and standard Antifers were monitored over a period of six years. Monitoring results concluded that the ECO Armor Blocks had significantly higher species richness and biodiversity of both invertebrates and fish compared to standard "gray" Antifer units.

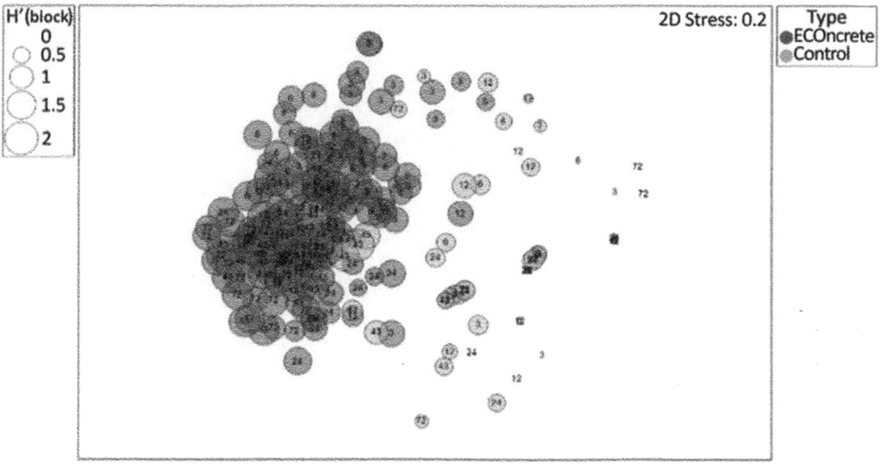

Figure 4.27. *Locations of ECO Armor Blocks and standard blocks on the dike (Perkol-Finkel)*

The diverse taxa assemblage seen on the ECO Armor Blocks included various types of engineering species, like oysters, tube worms and barnacles, that by deposition of a calcium carbonate layer on the concrete surface not only increase the availability and heterogeneity of biological niches on the structure but also contribute to the structures' stability and longevity via bioprotection, a process by which animals and plants protect the surfaces they colonize from weathering and erosion.

In addition, the ECO Armor Blocks presented a reduced dominance of nuisance and invasive species, as they opened the substrate to competition by native species. As soon as two years post-deployment, the ECO Armor Blocks had 21 native and 12 invasive species, whereas the standard Antifers had five native and 10 invasive species (Figure 4.28).

In conclusion, ecological engineering can be successfully applied into ports and working waterfronts, helping generate healthier urban marine environments that are more resilient and adaptive to climate change impacts.

Figure 4.28. *Comparative colonization of standard "control" Antifer blocks (left) and ECO Armor Blocks (right) designed by the company ECOncrete® (photo: S. Perkol-Finkel)*

4.4.3. *Landscape redevelopment of the East Darling Harbour waterfront in Sydney*[13]

The redevelopment of East Darling Harbour into Barangaroo is one of the most significant foreshore construction projects to date in Sydney, Australia. The entire land area represents an urban renewal project of 22 ha in total along 2.2 km of foreshore with the area divided into three precincts – South Barangaroo, Central Barangaroo and Headland Park.

Development has been occurring in stages since 2010, starting with Headland Park (the northern precinct). Headland park is a highly ambitious project that used the natural local substrate – sandstone – to re-create an entire foreshore area that follows the natural Sydney fault line (roughly 20 degrees north west) of Sydney Harbour, therefore crafting the new headland perfectly in line with natural Sydney Harbour headlands.

With Headland Park including more than 50% (1.4 km) of the foreshore within the development, this created considerable opportunity for the developers to enhance the intertidal habitat available at the site and recognize the cultural importance of the

13 Section written by Katherine A Dafforn, Senior Lecturer, Sydney Institute of Marine Science, Macquarie University, and University of New South Wales.

area. The total value of the Barangaroo development is estimated to be more than AUD$6 billion with AUD$250 million allocated to the design and construction of the 6 ha Headland Park.

The history of the site has been fairly mixed, but it was once part of the hunting and fishing grounds for the traditional owners of the land, the Gadigal people of the Eora Nation. The name "Barangaroo" is associated with a Cammeraygal woman of the same name. In the 1800s, the site was a hub of industrial activity and became known locally as the "Hungry Mile". Until the early 1900s, the site was occupied by a gasworks and then more recently by a container terminal until 2005, but it went into general decline after a new port was built to the south of Sydney Harbour in Botany Bay.

The intertidal area along the entire foreshore consisted of artificial, concrete vertical seawalls and was not accessible for the public. Plans for redevelopment of the site were announced in 2003 and construction began in 2010 with the entire site to be completed by 2021. Headland Park was the first precinct to be completed and opened to the public in December 2015.

Headland Park provides an interesting example of eco-friendly foreshore construction. The overall objectives for the site include the provision of public recreational space as well as the restoration of the headland to a semblance of its pre-1986 appearance. During the planning stages, consultation was done with various marine experts, including Danny Wiecek from the Department of Environment and Climate Change (DECC) NSW and Tom Heath from Kogarah Council. Guidelines such as those laid out in the "Ecologically Friendly Seawalls" document prepared by DECC NSW (now NSW Department of Planning, Industry and Environment) informed the design and construction of the new headland.

The final plans for the intertidal zone included replacement of a caisson seawall with natural coves and construction of a sloping seawall with 9,315 sandstone blocks sourced from the site. These sandstone blocks were cut individually from the rock escarpment with size, shape and placement along the foreshore pre-determined through computer modeling. Sandstone was selected as the material for the seawall because this was the local rock type found on natural shores in Sydney Harbour. The material has already weathered, with small rock pools forming and supporting some filamentous green and brown algae as well as coralline algae at lower intertidal levels (Figure 4.29). Placement of the sandstone blocks also created crevice refuges for diverse marine life, including oysters, snails, limpets and chitons.

Figure 4.29. *Headland Park seawall and detail of a basin created in sandstone, illustrating biomimicry and block aesthetics (photos: K. Dafforn)*

4.4.4. *CLI's experience with concrete dike shells: the need for eco-design*[14]

Artelia is a major historical player in the design of marine structures, including harbor dikes made of artificial block shells, through its subsidiary Concrete Layer Innovations (CLI).

Since its inception, Artelia has been committed to advancing its methods, tools, technologies and solutions for both its public and private clients.

14 Section written by Cyril Giraudel, Étienne Cunge, Michel Fons and Osanne Paireau, marine civil engineers and specialists in concrete shells from Concrete Layer Innovations (CLI), Artelia Group.

It has been innovating relentlessly for more than 50 years to keep pace with changes in the technical, environmental and societal expectations relating to these marine infrastructures, whose primary purpose is to protect the coastline and coastal developments. This is evidenced by the successive patents filed by the Group (Sogreah became Artelia in 2010): (i) the Tetrapod block for bi-layer shells, 1953; (ii) the Accropode™ block for single-layer shells, 1981; (iii) the Ecopode™ block, eco-designed for single-layer shells, 1996; (iv) the Accroberm™ block, a toe unit, an eco-designed version of which exists, 2018.

Ecopode™ blocks (Figure 4.30) were the first shell blocks eco-designed by Artelia and aimed to provide a concrete response to an important environmental and social objective of the 1990s: the landscape implementation of marine structures, particularly useful in emblematic places or remarkable sites. With their more natural texture and using concretes coated in colors adapted to local materials, they enable the emerged part of the infrastructure to be better integrated into its immediate landscape environment.

Today, within an environmental and societal context marked by climate change, the decrease in both marine and terrestrial biodiversity and the increasing phenomenon of rising sea levels, marine defense structures are considered as strategic infrastructures for the future.

On the one hand, given the context of climate change, the need to resort to such infrastructures is often unavoidable in certain areas, whereas more natural approaches (strategic withdrawal, nature-based solutions) represent approaches to be favored.

Figure 4.30. *Ospedaletti Ecopode™ and Accropode™II. In this project, the Ecopodes™ are coated to match the local landscape context (photo: C. Giraudel)*

On the other hand, experimental feedback highlights a strong and close link between marine structures and biodiversity. These interactions are of various natures and are on two distinct levels:

– the role played by the blocks forming these structures in colonization support, in the implantation and the development of marine fauna and flora;

– the overall impact of the structure in interfering with ecological continuities, which are essential to the proper functioning of coastal ecosystems.

Since 2012, Artelia has organized colonization monitoring of certain port dikes, more than 10 years old and constructed with boulders (single-layer shell), in order to better understand their colonization process (Figure 4.31).

The challenge now is to design and promote marine structures with increased biodiversity. To do this, infrastructures must both meet the perennial objective of protecting coasts against swells and, as much as possible, not degrade the richness and productivity of local ecosystems; or, better yet, they should play a role in preserving, restoring or even enriching the surrounding environment.

Figure 4.31. *Examples of colonization of artificial blocks of embankment shells observed during monitoring dives (photos: C. Giraudel)*

4.4.5. *Eco-design of the Brest harbor breakwater*[15]

At the crossroads of the Channel and Atlantic ocean spaces, with free access to the sea, the port of Brest is at the heart of European transport networks. In 2010, the Brittany region, which has owned the port since 2007, launched an ambitious development project to increase its competitiveness.

15 Section written by Anne Rioux, hydraulic engineer and project manager from Safege, Suez Environnement company, on the eco-design of the port of Brest.

The project includes, in particular for 2019–2021, the construction of an 860 m long enclosure dike and a 360 m long quay at the eastern end of the port. The enclosure dike consists of a gabionade of flat sheet piles embedded in an embankment, itself protected by a rock shell on the sea side. Ecoblocks are integrated into this shell to create artificial intertidal basins.

In their natural state, troughs are a special type of intertidal rock habitat and occur where the shoreline topography enables water retention in bedrock depressions. Here, the flora and fauna are permanently submerged, allowing subtidal species to survive. This habitat therefore corresponds to ecological enclaves. The deeper the basin (from several decimeters to a meter), the more it favors biodiversity.

The development project for the port of Brest was the subject of numerous environmental assessments between 2011 and 2014, which contributed to adjusting the work program in accordance with the "Avoid, Reduce, Offset" approach. In 2014, as part of its mission to prepare the project's impact study, Suez Consulting, in a note on the opportunities for ecological enhancement of marine structures, suggested the first outlines of the eco-design of the dike, with the proposal to create mediolittoral basins. The Brittany region adopted this objective of increasing the ecological value of the infrastructures, which has translated into accompanying measures for the project.

Bouygues TP Régions France, holder of the contract for the dike, along with Setec In Vivo, proposed in 2017 the manufacture of a large cubic ecoblock that would create a deep basin and correspond to a habitat that would promote maximum biodiversity. The eco-design of the dike includes no less than a hundred ecoblocks that are integrated into the shell of the dike along its entire length.

In concrete terms, the ecoblock (Figure 4.32) is a 1.9 m square cubic block of 1.9 m, partly hollow, with dimensions equivalent to those of the largest rockfill used to build the shell of the dike so that it can be integrated into the profile of the structure without modifying its stability. The basin is delimited on the inside by a pyramidal concrete mass, the surface of which is made from deactivated concrete in order to create roughness. Once the block is placed diagonally in the shell and inclined at a 3/2 slope, the basin is filled through the open face of the ecoblock according to the tides.

For the Brittany region, this operation to improve the ecological value of the infrastructure is providing strong feedback that will benefit other regional ports. From this point of view, the monitoring of the effectiveness of the accompanying measure will be defined and then ensured by the IUEM-UBO "Fauna Flora" observatory. The IUEM is a member of the scientific select committee set up by the Brittany region to monitor the work.

Figure 4.32. *View of the intertidal basins on the dike (photos: P. Hymery, Suez Consulting)*

4.5. Biomimetic artificial reefs in Corsica (Ajaccio)

The Corsican Environmental Office (*Office de l'Environnement de la Corse*, OEC), the contracting authority, has undertaken an operation for new-generation structures at the bottom of the Gulf of Ajaccio, at Ricantu beach (Figures 4.33 and 4.34). The location chosen by the OEC takes into account the different uses of the bay and has been the subject of wide consultation with local stakeholders. In addition, the location of the structures, which takes into account an ecosystem approach, is to the south of the sandy area of Ricantu, close to rocky areas and Posidonia meadows, at the southern end of the bay.

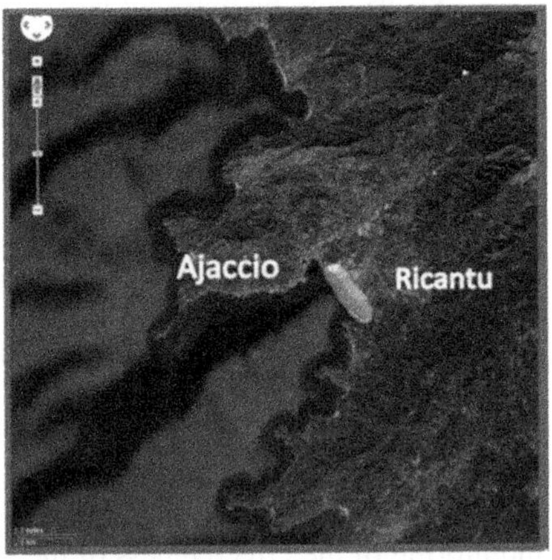

Figure 4.33. *Location of the immersion site for the eco-designed artificial reefs structures, Ricantu Bay, east of Ajaccio Bay (source: OEC document, A. Agostini)*

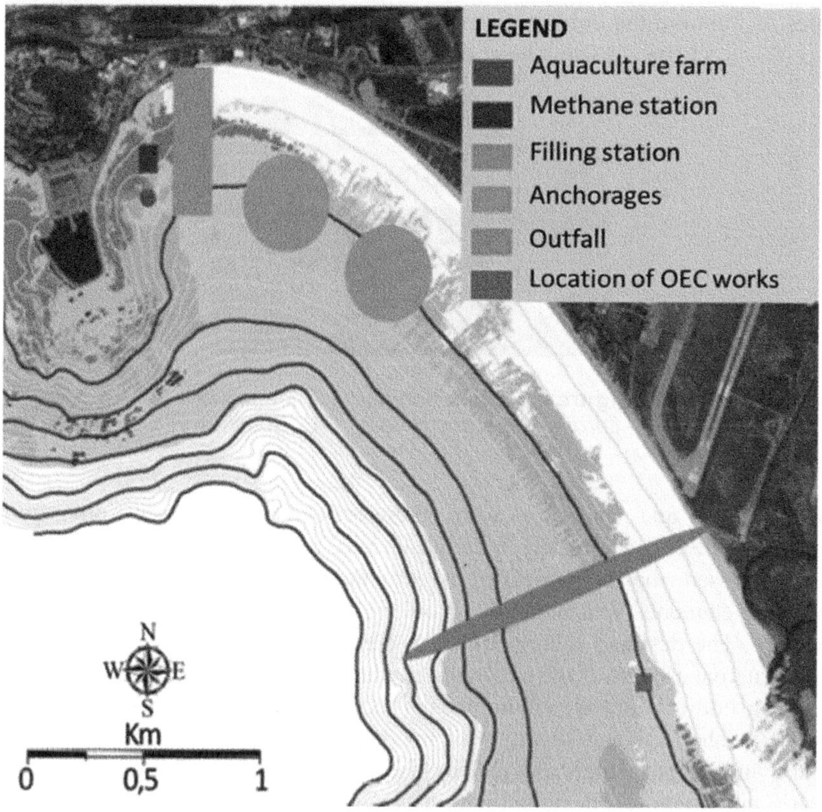

Figure 4.34. *Immersion sites for the structures (red square) at Ricantu bay (source: OEC – ©IGN 2015, A. Agostini). For a color version of this figure, see www.iste.co.uk/pioch/marine.zip*

Following consultation with local stakeholders, OEC concluded that an innovative, experimental, small-scale research approach was needed in order to create experimental structures, adapted to local conditions and to be monitored scientifically for three years. At the end of this period, the OEC plans to implement a large-scale project that will build on the results of the experimental phase.

The experimental structures consist of bio-inspired artificial reefs and ecologically designed experimental buoys (Figure 4.35). The approach adopted for the design of the structures must be inspired by the ethological knowledge of the local target species and the shape of the local natural habitats. This is the most significant innovation of this type of artificial reef with biomimetic objectives, achieved through the study of shapes as well as through the research of natural rock textures to improve the colonization of local marine organisms and the installation of specific target species (Souche *et al.* 2019).

Figure 4.35. *Eco-engineered mooring for cardinal beacons (photo: J.-C. Souche)*

The shapes and materials were therefore chosen to meet environmental specifications. Furthermore, given the absence of a scale effect, the economy of the project imposed the construction of structures for which the weight of the constituent elements was limited according to local capacity for handling and immersion. The designer therefore had to consider a modular design and integrate the constructive aspects of the execution methods from the start of the studies, while also taking into account the environmental requirements. At this stage, the support of snorkelers, fishermen and divers was invaluable in finalizing the design of the structure and, in particular, the species–habitat relationship. The result was unique experimental structures, designed with a systemic approach and the implementation of which was in line with local technical resources.

The designer therefore started with a blank page with a view to achieving the following objectives: (i) offering habitats adapted to the target species according to the required stages (juveniles and adults); (ii) restoring rocky seabeds degraded by human activities; (iii) testing new immersion concepts; (iv) proposing a design with an impact on benthic and pelagic target species; (v) testing new types of modular structures; and (vi) testing different types of materials: concrete, steel and natural bioconcretion, using the electrolysis system developed by the Geocorail© company.

Concrete as a material has offered design freedom for some structures since it can be freely molded and can be used to treat surface texturing. The structures for this project were manufactured by the company Isula Services, based in Ajaccio, using a concrete dosed with 350 kg of cement CEM I 52.5 MPa and with an E/C ratio = 0.55. The average strength values obtained at 28 days were in the order of 35–40 MPa (class XS2). The formwork of the elements was carried out using an elastomer mold supplied by LIB Industries. The use of this type of formwork enables a multi-scale porosity and roughness to be obtained, which is favorable to the development of fouling. This system is adapted to meet biomimicry objectives since it enables a surface state identical to a natural rock to be reproduced (Figure 4.36).

a) b)

c)

Figure 4.36. *Concrete elements cast by Isula Services: a) counter-mold; b) mold and reinforcement; c) demolding (photos: J.-C. Souche)*

This experiment, conducted by the OEC, enables the eco-design strategy that owners can adopt in the construction of artificial reefs and buoys to be explained. With regard to the material parameter, the project underlined the difficulty of working on all the parameters in parallel. A compromise between strength and durability requirements and the capacity for colonization by living organisms must emerge from an eco-design approach. The ecological objective of establishing a habitat–environment relationship does not allow the use of standard off-the-shelf solutions but requires the development of technical solutions specific to each project and each site. Finally, during the operation of the structure, ecological monitoring is essential to objectively evaluate the biological performance of the proposed technical solutions (Figure 4.37). Experience shows that it is possible to propose an eco-designed, economically viable solution, so long as it is proposed from the outset. It can even be argued that integration of all the constraints of this site within the framework of the eco-design approach qualified the financial feasibility of the operation since the design was completely adapted to the local technological means with the aim of economic feasibility as a priority.

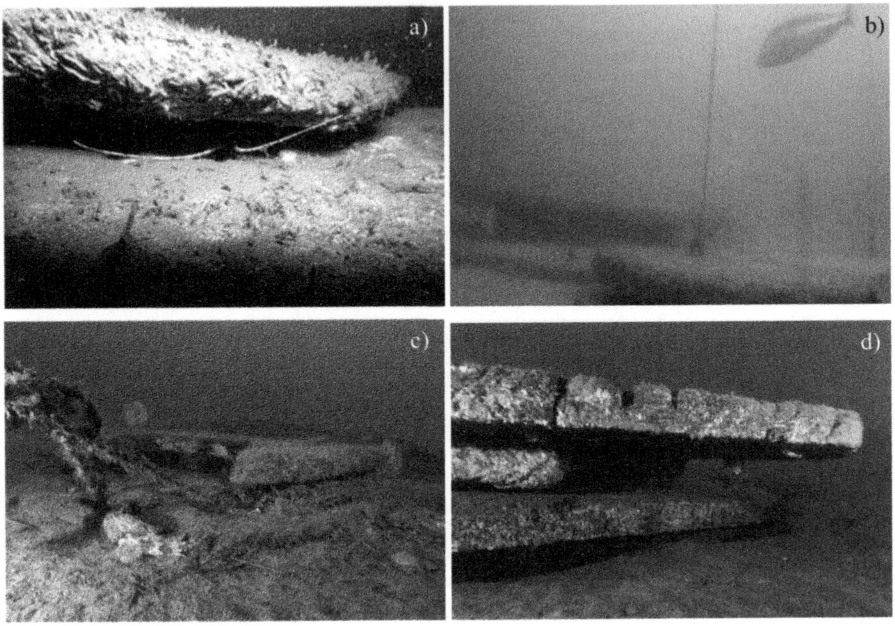

Figure 4.37. *Eco-designed artificial reefs and environmental monitoring (Gulf of Ajaccio, OEC), which validated the presence of target fishery species for Corsica: a) spiny lobster; b) dentex; c) red scorpionfish; d) biomimetic reefs (photos: A. Agostini)*

"Corsica possesses an exceptional environmental and landscape context. Our very well-preserved coastline is the result of a relentless fight over several decades to maintain this natural wealth.

The eco-design of developments is one of the avenues that should be explored in order to: (i) face the significant evolutions between now and 2030 (in terms of artificialization of the coastline and tourist frequentation), in order to maintain our environmental wealth within a pleasant living environment; (ii) refuse excessive artificialization and search for another, new kind of development, different from what has been done in regions that welcome mass tourism.

Indeed, marine structures can be aesthetic, integrated into underwater landscapes and useful to nature. The *Office de l'Environnement de la Corse* wished to be part of this approach: in 2017 through the implementation of new-generation artificial reefs of a biomimetic design, in 2018 through eco-designed buoys that were made available to the DIRM MED South Corsica Lighthouse and Beacon Service and, finally, by providing financial and technical support to the Ajaccio South Corsica Chamber of Commerce and Industry for the creation of eco-designed moorings dedicated to accommodating large and very large passenger ships (projects underway).

The OEC has initiated a project for the creation of new-generation artificial reefs, positioned on the seabed of the Gulf of Ajaccio, in the Natura 2000 zone (FR 9402017) on the Ricantu site. The objectives of this operation are threefold: fisheries production, reconstitution of marine biocenoses and research and development. The OEC wanted to be part of an R&D approach by innovating the architecture and texture of the modules, as well as the manufacturing techniques. The reflections were based on biomimicry and, in order to keep costs reasonable, the consideration of local transport and lifting resources.

Innovation being an axis of the project, it was not a question of importing existing reefs or 'ready-made' know-how, but rather of creating modules that had never been made before. Indeed, as a universal solution does not exist, it was necessary to create reefs that met the local needs of specific target species, 'made to measure'!

This innovative and experimental approach was based on numerous sets of feedback in this field (bibliographic syntheses), while taking into consideration local specificities thanks to the involvement of the concerned actors on the territory (professional fishermen, snorkelers, scientists, organizations, state services and communities). The designer listened to these actors and designed the structures with the objective of reproducing the forms of natural habitats used by species (of economic interest) present on this site. Indeed, the analysis of the site characteristics and the behavior of the fish (ethology) enabled the geometry and texture of the modules and their layout on the seabed to be developed. Furthermore, from the beginning of the project, the OEC envisaged the integrated management of this site, including regulation, ecological monitoring, surveillance, communication and public awareness.

The functions targeted by the establishment of artificial habitats are feeding, shelter and breeding areas, primarily for lobster, capon, sar and denti. Ecological monitoring (carried out in 2018 and 2019) has highlighted the accomplishment of these three functions. The initial results are very encouraging, with the presence of species with high market values, such as red scorpionfish (adult), white seabreams, dentis, juvenile spiny lobsters, forkbeards, John Dory, etc. (Figure 4.37). In order to complete this experimentation with three artificial reef (AR) prototypes, the OEC commissioned Isula Services to produce two eco-designed weights for offshore buoyancy made available to the Lighthouse and Beacon Service during the summer of 2018."

Box 4.3. *The genesis of the Ricantu reef project, Alexandra Agostini*[16]

16 Principal assistant in the species and natural environment department of the OEC, Ajaccio.

4.6. Artificial island eco-design

4.6.1. *The Monegasque government's vision for the Principality of Monaco's offshore extension*[17]

The Principality of Monaco is backed by a mountainous massif and faces the sea. A very attractive territory, its population is the most dense in the world, with 18,866 pop/km², which is more than double the next highest of 7,829 pop/km² in Singapore. It is, therefore, logical that extension of the territory by filling in the seabed has developed since the end of the 19th century (Figure 4.38). In the late 1800s, the first fillings were carried out at Larvotto (3 ha), Fontvieille (6 ha) and at the bottom of the Bay of Hercules, where the eponymous port is located (7.5 ha).

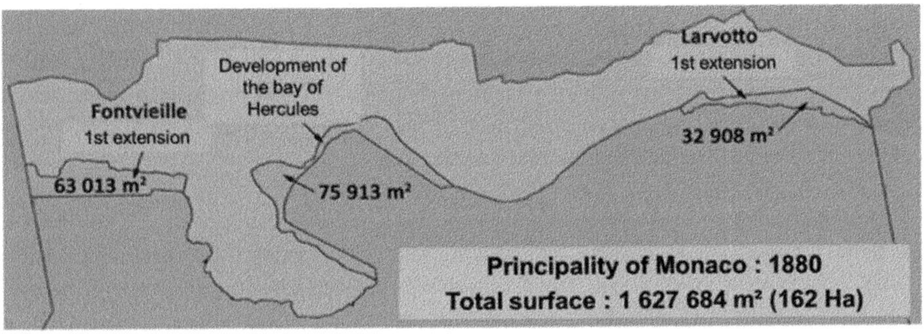

Figure 4.38. *Plan view of the Principality of Monaco in 1880 (source: Princely Government). For a color version of this figure, see www.iste.co.uk/pioch/marine.zip*

The 20th century was an even more active period in this field, with the development of the Larvotto and Portier reclamations in 1965–1970 (gain of 10 ha) and especially with the extension of Fontvieille (1970–1980) leading to a gain of an additional 22 ha (Figure 4.39).

In total, about 40 ha, or 20% of the territory, have been reclaimed from the sea in the last 150 years.

In 2013, S.A.S. Prince Albert II announced the launch of a new extension project (peninsula) in the Anse du Portier district, with a strong environmental ambition.

17 Section written by Jean-Luc Nguyen, director of the urbanization at sea mission, Principality of Monaco.

This project, for which the ecological stakes are off the charts, will enable the creation of a residential area (luxury collective and individual housing) with shops (about 3,000 m^2), public spaces offering a wide variety of atmospheres, a central square and a landscaped park, as well as public infrastructures (including a parking lot, an entertainment port (Figure 4.41) and an extension of the Grimaldi Forum of about 6,000 m^2).

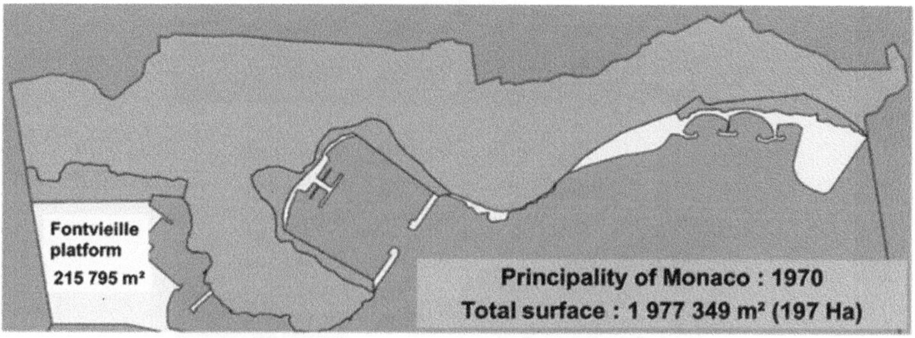

Figure 4.39. *Plan view of the Principality of Monaco in 1970 (source: Princely Government). For a color version of this figure, see www.iste.co.uk/pioch/marine.zip*

a) b)

Figure 4.40. *Artist's plan view of a) the project and b) its waterfront (source: Princely Government)*

From the outset, the environmental objectives of this project were very ambitious, as the new extension is limited to the northeast by the Larvotto marine reserve and to the southwest by the protected area of the Spélugues coral reef (Figure 4.42).

a)

b)

Figure 4.41. *Artist's view of a) the harbor and b) the urban context (source: Princely Government)*

Figure 4.42. *Layout of the offshore extension of the Portier cove (light blue hashed area) between the Larvotto marine reserve and the Spélugues coral reef (source: Princely Government)*

As part of the environmental impact study, a very precise initial assessment was carried out. Sensitive areas and heritage species were identified and safeguard measures were implemented (relocation of protected species).

The design of the project took into consideration the limitation of impacts during the works as well as in the final phase; for example, with regard to the circulation of currents. The private group in charge of the works was also tasked with carrying out environmental monitoring of the project over a 10 year period; this monitoring will later be carried out annually by the Principality.

In order to choose the technique for the construction of the marine infrastructure, the impact reduction measures were analyzed. Once the solution was chosen, that is, an embankment confined by a belt of reinforced concrete caissons, the company was committed to construction methods adapted to minimize damage to the marine environment, with particular attention to controlling turbidity and sedimentation.

The main environmental measures taken during the construction phase were:

– the use of equipment adapted to the context: ecological skips for sediment dredging, dredging without overflow;

– the installation of several hundred meters of underwater screen to protect the neighboring nature reserves;

– continuous monitoring of turbidity and sedimentation (monitoring device with real-time sensors and alert thresholds);

– the movement of protected species.

Compensation measures have also been defined:

– a program of cleaning and restoration of the Spélugues reef;

– eco-design of the caissons;

– installation of targeted artificial reefs.

To accompany the project on the environmental side, the state set up an environmental monitoring committee, made up of independent experts. During the study phase, the role of this committee was to examine the measures intended to protect the environment and to formulate proposals for their improvement. During the construction phase, this committee ensured that the environmental measures are effectively implemented, with a view to continuous improvement.

Under the direction of the environmental monitoring committee, the accompanying measures and environmental integration of the maritime infrastructure significantly evolved:

– systematic treatment of the walls of the caissons by grooving or die-cutting and addition of ECOncrete® modules (Figure 4.43) and bioconcretionary nets was decided;

– wave damping chambers were eco-designed, with the insertion of artificial habitats (reefs, cavities, internal surface matting) (Figures 4.44 and 4.45);

– ecological continuities were created in the infrastructure embankments, towards an off-project artificial reef zone.

Thus, 17 mega-caissons, each one 10,000 tons of concrete and steel, were eco-designed thanks to a partnership between the project owner SAM l'Anse du Portier, the constructor Bouygues TP, the Princely Government services and a group of international experts in the field of eco-design.

It is important to note that the principality opted for developments that included an ecological function from the design phase.

The Monaco offshore extension project illustrates the fact that biodiversity is a priority that can be perfectly reconciled with economic and urban development, if the resources are made available to impose, right from the design phases, the consideration of precise ecological functionalities, and to reach a cost/ecological efficiency ratio that is acceptable for both nature and humankind. Through this iconic structure, the Principality of Monaco is contributing to the progress of eco-design standards. We hope that future structures in contact with the sea will be inspired by this approach.

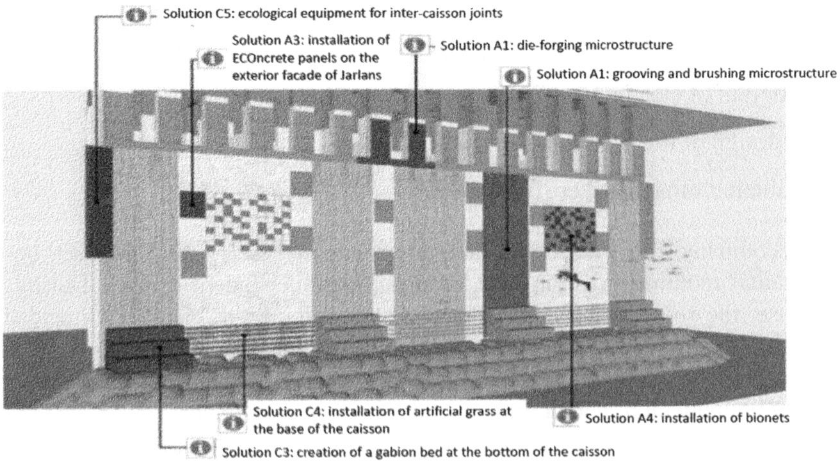

Figure 4.43. *Elevation of the caissons with the planned eco-design provisions (source: Princely Government)*

Figure 4.44. *Details of the "Amur" die-cutting on the posts of the Jarlan caissons (source: Princely Government)*

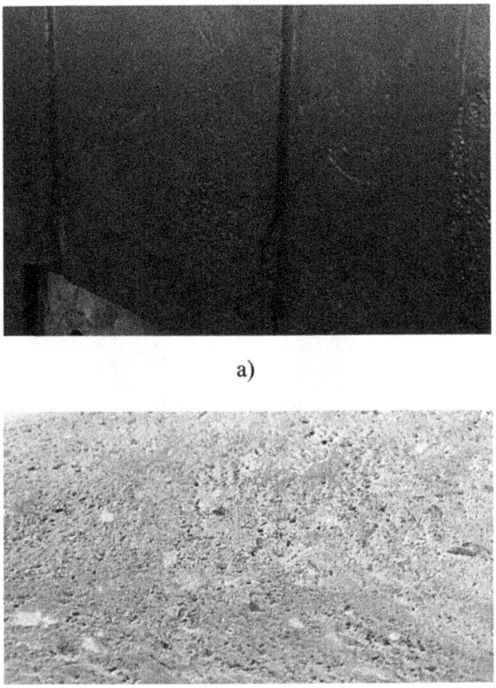

a)

b)

Figure 4.45. *a) Grooving of the caissons, and b) sanded concrete surfaces (source: Princely Government)*

4.6.2. *The genesis of the eco-design of the Monegasque artificial peninsula project*

In the context of a previous reclamation project, initiated in 2007, an eco-design approach was proposed to the Monegasque government by one of the candidates. The project aimed to create 15 ha of artificial soil, with foundations installed down to 50 m below the seabed. The eco-design of this colossal structure was particularly complex due to the requirement to create small rocky seabeds for juveniles. We simply present here the eco-design orientations that were taken and the ecological principles that governed their development, to illustrate the originality of the approach.

The impacts on the seabed were highly significant. The eco-design focused on the artificial island walls and the creation of connectivity through the placement of adjacent artificial reefs, enabling improved biological flows between three major biological reservoirs (denoted as R1, R2 and R3 in Figure 4.46).

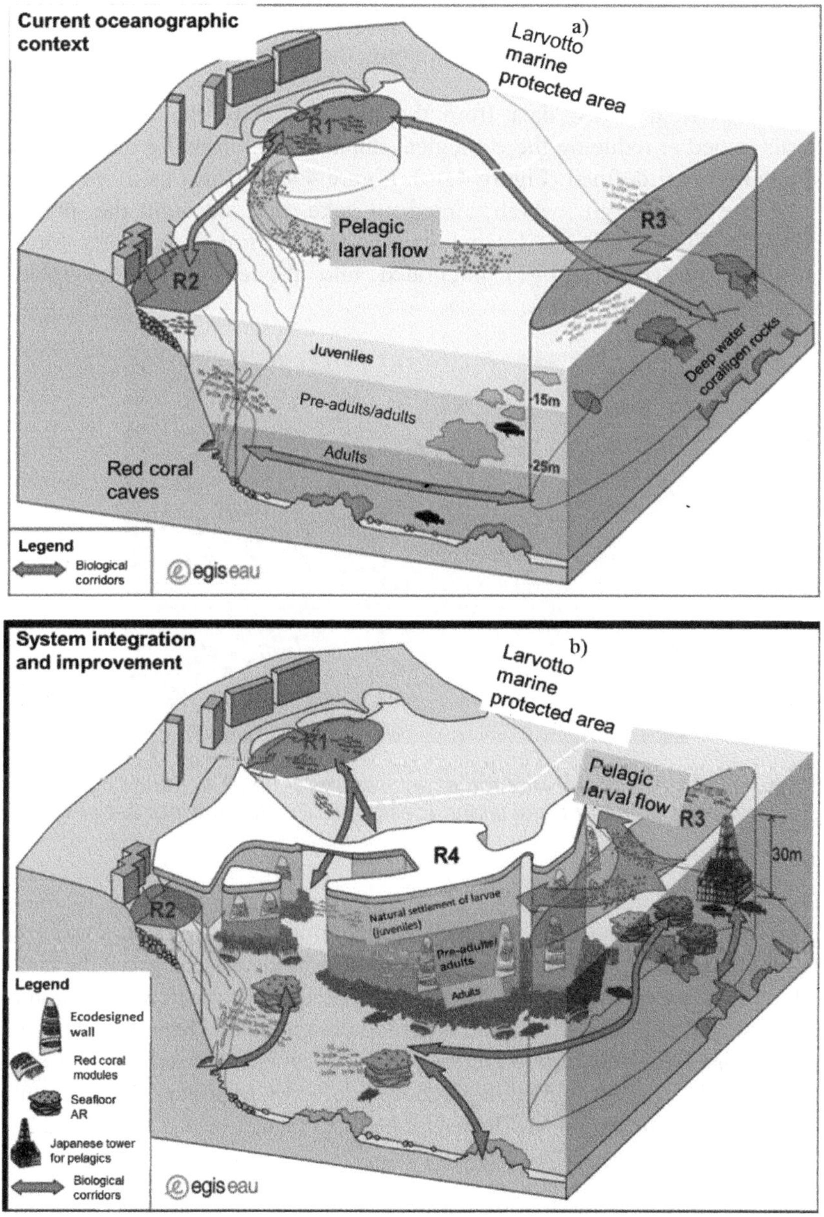

Figure 4.46. *a) Diagram of the functioning of the ecosystem surrounding the artificial peninsula project; b) diagram describing how the eco-designed structure can interact with this environment, reducing the negative impact (design: S. Pioch, drawing: J.-L. Feron, Egis Eau, 2008)*

An initial map, schematizing the functioning of the ecosystem surrounding the project, was proposed (we reproduce here only the general layouts).

It is from field dives, data from the initial state of the impact study, that proposals aimed at reducing the ecological impact and improving the integration of the structures were defined (Figure 4.46). Figure 4.46 enables us to understand the logic of the eco-design, which aimed to take into account the pre-existing connectivity (Figure 4.46(a)) by creating corridors through supports for marine fauna and flora (Figure 4.47) integrated into the design of the underwater infrastructures (Figure 4.46(b)).

Figure 4.47. *Artist's view of the structures integrated into the design of the foundations of the artificial island (design: S. Pioch, drawing: J.-L. Feron)*

This 15 ha extension project did not see the light of day due to the global financial crisis of 2008 and because of lingering doubts about the control of ecological impacts on the Larvotto reserve adjacent to the project, given the circulation of bodies of water and the Monegasque seabed. Studies for the new extension project were relaunched in 2015, with an architectural design led by Denis Valode and Renzo Piano, associates of landscape architect Michel Desvigne, based on 60,000 m^2 of reclaimed land, an area deliberately reduced in order to lessen the environmental impact (see the testimony of Mr. Jean-Luc Nguyen in section 4.6.1).

4.7. Eco-design of mooring systems

During the summer season, pleasure boats and large yachts are very common in ecologically rich and sensitive areas, such as the Mediterranean seagrass beds (Posidonia meadows in particular) or coral areas in the French West Indies. These vessels are moored using "plow" anchors that turn over the seabed and destroy seagrass beds, Mediterranean coralligenous zones and corals. The public authorities have therefore decided to organize the mooring of these vessels in order to protect these sensitive areas. This new approach involves the establishment and development of mooring zones, which, depending on the sites and needs, could be subject to an eco-design approach.

4.7.1. *Eco-design of moorings for recreational vessels in Guadeloupe*

The French West Indian seabed is made up of coral and seagrass beds. These ecosystems are highly vulnerable to the increase in pleasure boating, since moorings cause serious destruction (repeated anchorage) in the most beautiful tourist bays. In the long term, in a few years, these seabeds will be destroyed, causing losses of biodiversity and important economic losses. Under the direction of the *Société d'Economie Mixte de Saint-Martin*[18] (SEMSAMAR), a vast program of organized mooring implementation has been deployed over the last 10 years; it is led by Vincent Derussy.

4.7.1.1. *A first experience in Deshaies in 2013*

Each year, as a result of unauthorized anchoring, considerable damage is caused in the magnificent tourist bay of Deshaies (Figure 4.48).

Studies carried out by the *Direction de l'Environnement de l'Aménagement et du Logement* (DEAL) in Guadeloupe and Martinique show that controlling the impact of unauthorized anchoring is a priority objective for the conservation of the coral seabed on the Caribbean coast (Pérès 2006).

In spite of numerous recommendations, or even prohibitions, the only effective measure to avoid this damage and maintain vital tourist economic activity are organized moorings. These classically consist of buoys, metal plates sealed to the rock or screws anchored in the sand (Pioch and Leocadie 2017).

Within the framework of the implementation of a mooring and light equipment zone (*Zone de Mouillages et d'Equipements Légers*, ZMEL) in 2013, the commune of Deshaies, in the north-west of Guadeloupe, requested an eco-design component, on the advice of SEMSAMAR, in order to preserve its remarkable coral seabeds and to offer shelter to juveniles against the lionfish (*Pterois volitans*, an invasive predatory species in the Caribbean since the 1990s).

---

18 Mixed economy corporation, public/private partnership of technical and financial support for the development of local authorities of the French overseas territories of Saint-Martin, Guadeloupe, Guyana and Martinique.

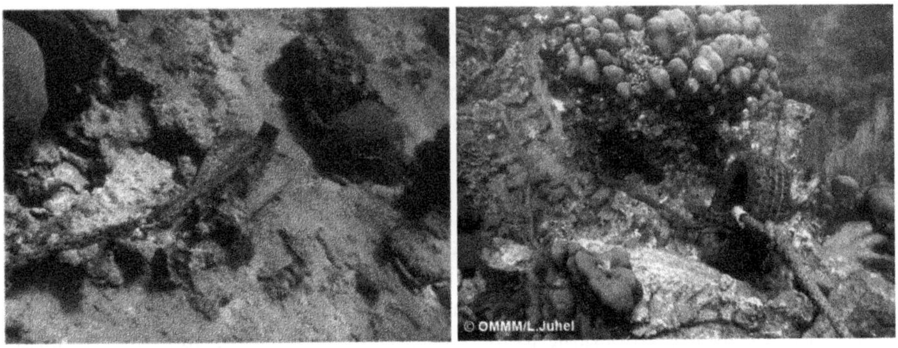

a) b)

Figure 4.48. *a) Effects of anchors on the seabed of the Caribbean coast; b) wild moorings installed by default (source: DEAL – OMMM, L. Juhel, 2006)*

It was an opportunity to "kill two birds with one stone", in the words of the mayor. Here again, the study of local species underwater, the study of the behavior of juveniles in natural habitats and exchanges with local fishermen enabled us, with engineers specialized in marine civil engineering from the company Suez Environnement, to develop several models of eco-designed buoys.

Observation of natural environments and discussions with fishermen protecting juvenile lionfish enabled us to mimic certain habitats that seemed conducive to protecting juveniles (corals arranged in vertical stalks called "deer horns" and mangrove roots) by creating "skirts" surrounding the buoy (Figures 4.49 and 4.50).

Four years after immersion in 2014, ecological monitoring of the 45 eco-designed modules provided interesting results, both in terms of the juvenile colonization of these structures and their protective role against lionfish over-predation (Figure 4.51).

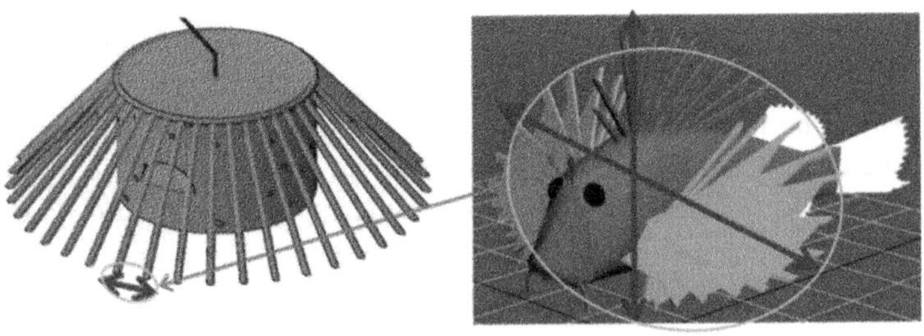

Figure 4.49. *Natural habitat-inspired skirts attached to buoys to prevent access by predatory lionfish to the juveniles and create refuges in cavities developed in the buoys (S. Pioch)*

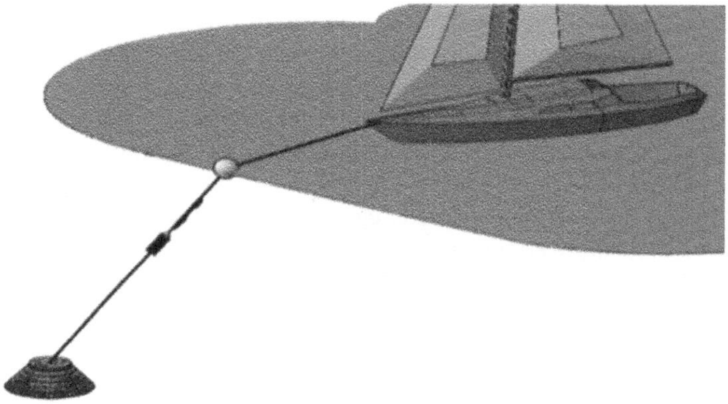

Figure 4.50. *Eco-designed mooring system (design: S. Pioch, drawing: J.-C. Ascione, Safège SAS)*

Figure 4.51. *Juvenile colonization of an ecologically designed mooring system after five years of immersion (photo: S. Pioch, 2018)*

Reproduced below are the conclusions of the report by Claude Bouchon, Yolande Bouchon-Navaro, Samantha de Lavigne and Sébastien Cordonnier from the University of the West Indies, the Borea laboratory and the Caribbean consulting firm Aqua Conseil, respectively:

> In concluding this study, it appears undeniable that the "ecological" modules constituting the buoys exert an attractive force on benthic

organisms and fish, since 58 species of benthic organisms and 43 species of fish were observed on and around the buoys. In comparison, 74 species of benthic organisms and 25 species of fish were recorded on the coral mass near the buoys, the surface area of which is much larger than that represented by the 22 buoys studied.

Statistical analysis of the results clearly showed that both benthic and fish biodiversity were related to the architecture of the modules and the level of complexity. The most attractive module types for benthic animals, and especially for fish, were the simple blocks and those with a double tray associated with a belt of PTFE tubes.

It is possible to separately examine the influence of the diverse buoywork installed on the modules. The basic buoywork of all the modules, i.e., 6 cm diameter holes, exerts a definite attraction on the crustaceans and, in particular, on the young lobsters. On the other hand, the half-moon shaped tunnels are often empty and do not seem to present any obvious interest for the animals.

Rocks of natural origin, of decimetric size, have been glued to the upper surface of certain modules in order to facilitate the installation of benthic organisms. It is true that the larvae of these organisms, and especially the corals, avoid horizontal surfaces during their recruitment and prefer sloping or even vertical substrates. However, rather than the nature of the substrate, it is its occupation by a complex biofilm which prevails in their fixation. In order to simplify the construction of the modules, these rocks could be replaced, at the time of the casting of the blocks, with the surface irregularities of concrete, made in a mold. These could be truncated and conical, pyramidal or cylindrical in shape.

The upper part of some of the modules supports a circular concrete tray, 8 cm from the main block. This structure is very attractive for fish and, to a lesser extent, lobsters. Some simple concrete blocks, or those with a double top plate, have been equipped with a belt made of plastic tubes (PTFE). This is certainly the most attractive equipment for fish. Moreover, they have been largely colonized by fixed benthic animals which found them to be a favorable substrate. This equipment has resisted perfectly against the swell from Hurricane Maria. Furthermore, four years after the installation of these structures, examination of the PTFE tubes has not indicated any apparent sign of chemical corrosion or significant abrasion that could pose a risk of plastic contamination of the surrounding environment.

It is interesting to note that no lionfish were counted inside the “skirts” imitating mangrove roots, intended to offer protection to juveniles, during the three monitoring dives organized since the installation of the structures. In comparison, the old buoys left in place, made of tires filled with concrete, did not offer any shelter to the local ichthyofauna.

The additional cost of this operation of about 10% cannot be considered negligible, since the materials used and the innovative aspect have forced the manufacturing company to test several systems in order to validate the one that will, in the future, produce the best ratio between durability and price.

4.7.1.2. *Eco-design of moorings in Bouillante (2018–2020)*

As part of the development of its tourism potential and the capitalization of its natural maritime assets, the commune of Bouillante plans to create a light cruiser stop (*Halte Légère de Plaisance*, HLP) using eco-designed moorings on the three sites of Bouillante, Malendure and Anse à la Barque (Figure 4.52).

In order to develop a technical solution that takes into account all the constraints of these three sites, SEMSAMAR, which led the operation, commissioned the companies BRL Ingénierie, Urbacité and LTC to set up ecologically designed buoys for pleasure boats of about 15 m, based on the successful Deshaies experience (2003).

The sensitive and patrimonial character of the site has led the municipality to set up light infrastructures with high environmental value, which are integrated into the natural environment. In addition to preserving the seabed, the commune also intends to revitalize the local marine fishing communities. A total of 100 eco-designed moorings will be distributed over three sites along the coast.

This project targets local economic development, in partnership with local actors, as well as some ambitious ecological objectives defined by Caraïbes Aqua Conseil:

– recreate a nursery function: habitats for young lobsters and other local juvenile or pre-adult benthic coral species (snappers, parrotfish, pomacentridae, etc.);

– create a nursery function: establish fauna and flora allowing the development of the nursery function in the water column and on the seabed (hard substrate);

– capture pelagic and demerso-benthic larvae using buoys located on the surface (mooring);

– develop hard substrates to restore coral beds and associated species (improve ecological resilience): rough and biomimetic substrates (surfaces mimicking the roughness of local rocks) to facilitate the establishment of new organisms from adjacent coral reservoir populations;

– improve landscape connectivity: create coral corridors and ensure better landscape integration;

– combat over-predation of juveniles by lionfish: the biomimetic concept consists of taking inspiration from the roots of mangroves where the "prison bar" structure does not allow lionfish to pass because of their fins (Figure 4.53). In natural mangroves, this type of structure protects juvenile fish from pre-destruction, as they can take shelter behind the artificial vertical curtain (as observed by local fishermen).

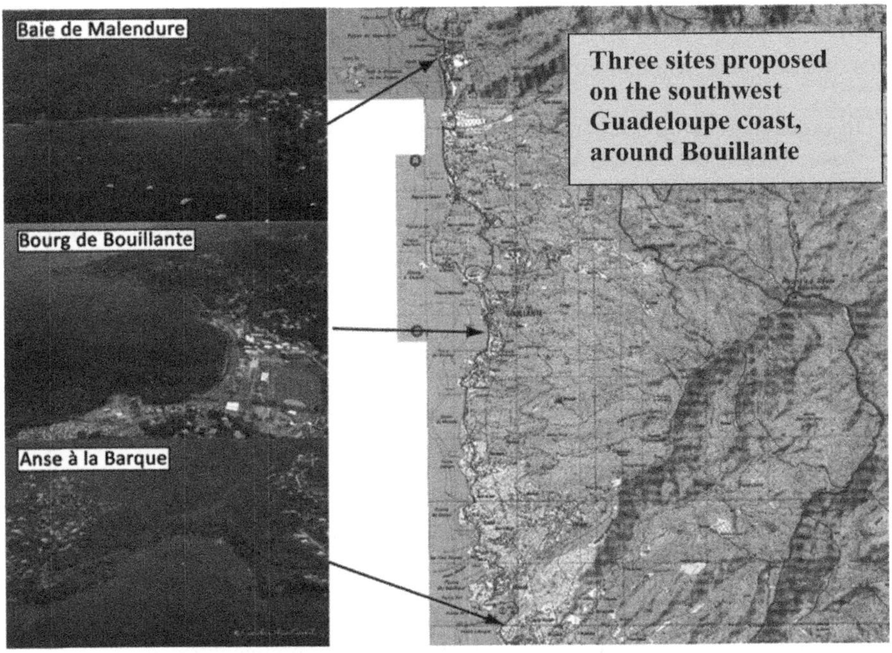

Figure 4.52. *Location of the three sites (document: S. Delavigne, source: Caraïbes Aqua Conseils 2017)*

Finally, the feedback from Deshaies has enabled us to improve the system and to develop new techniques.

For example, the calculation of stabilities with respect to cyclonic swells led to a significant increase in the tonnage of the buoys and to provision for a slight silting of the base in order to promote a suction effect; the peripheral skirts that protect juveniles from lionfish were enlarged and the reservations were modified (size, depth, number, etc.) specifically to take into account the recommendations from the monitoring carried out on the Deshaies structures.

a) b)

Figure 4.53. *a) "Prison bar" effect of stilt roots of red mangrove (*Rhizophora mangle*); b) span of a lionfish (photos: S. Pioch)*

The shape of the ballast was also revised to fit between a cylinder and a rectangular parallelepiped. However, the main innovation consisted of reproducing the very rough texture of local rocks to meet the objective of establishing living organisms (particularly adapted to coral larvae).

This surface condition was obtained by placing an elastomer skin manufactured by LIB Industries[19] into casings from the company Sogetras[20], holder of the work contract. It was completed by using different materials in accordance with the shapes and spaces adapted to the target species. The mooring line was also equipped with elements allowing the capture of larvae (Figure 4.54).

We can recall, in passing, that one particularity of eco-design is that it is influenced by successful experiences, by following natural evolution. The designer is always searching for the triple point between the right technicality, the expected ecological results and all this at the best cost[21].

19 http://www.groupelib.com/.

20 https://www.sogetras.fr/.

21 The authors wish to express their sincere thanks to Samantha Delavigne of Caraïbes Aqua Conseils, Romain Danlos of BRL Ingénierie and Vincent Derussy of SEMSAMAR: http://www.semsamar.fr/.

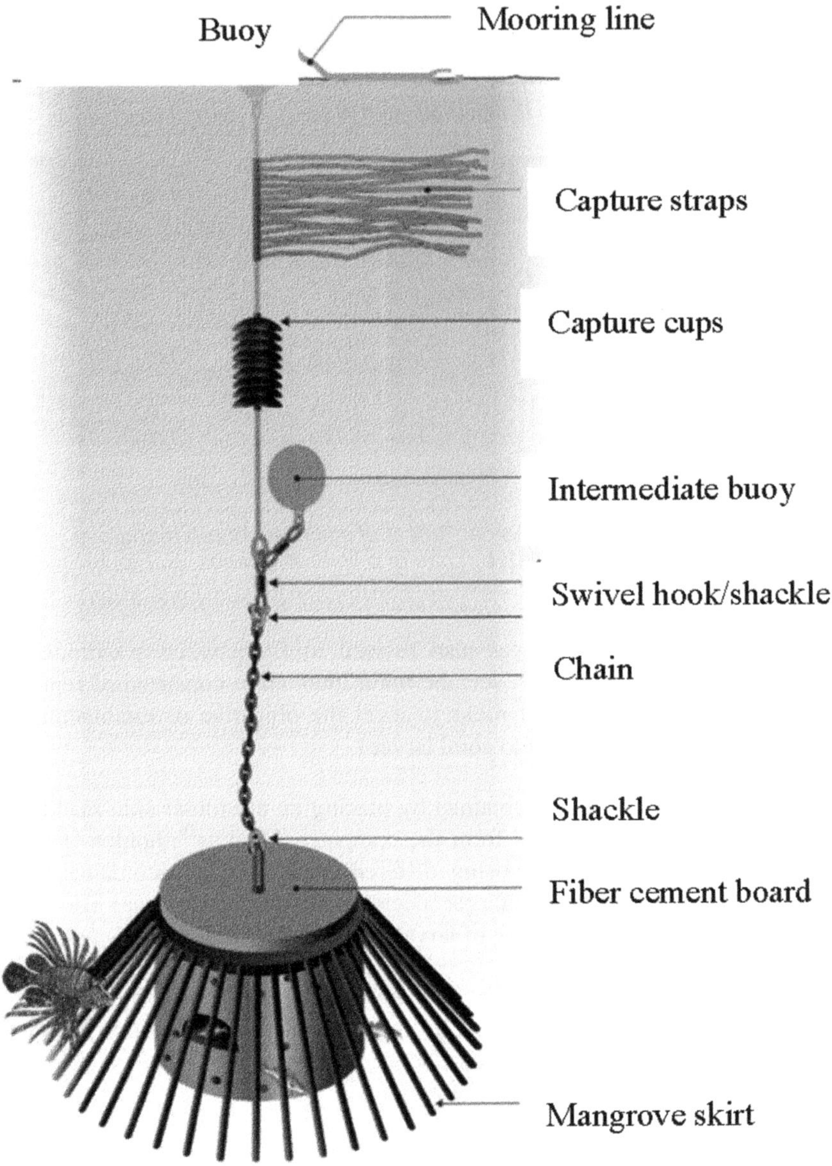

Figure 4.54. *Eco-designed mooring system, artist's view (design: S. Pioch and J.-C. Souche, drawing: R. Danlos, BRL Ingénierie)*

Figure 4.55. *Mooring system for the HLP Bouillante operation in Guadeloupe (photos: J.-C. Souche)*

4.7.2. *Eco-design by moorings for large vessels and yachts as a solution for the protection of the Mediterranean Posidonia meadow*

4.7.2.1. *Actions taken by the Maritime Prefecture of the Mediterranean to protect the Posidonia meadow*[22]

Posidonia (*Posidonia oceanica*) is a protected Mediterranean species. It covers 34% of the seabed along the French coast and up to 66% of the seabed of Corsica, between 0 and 40 m.

Posidonia meadows are extremely rich and vital ecosystems for many species. They support many ecosystem services from which humans benefit, but they are at risk from a growing threat: ship mooring. The seagrass beds have lost 10% of their surface area over the last 100 years, and the loss of services is estimated at more than 4 billion euros per year. In some tourist areas in the Provence-Alpes-Côte d'Azur region, degradation is even more significant due to stronger pressures, putting local authorities in the difficult position of having to choose whether to protect the seagrass beds or promote the local economy.

22 Section written by André Grosset, deputy head of the Protection and Sustainable Development of the Marine Area Division, Maritime Prefecture of the Mediterranean.

Faced with this observation, state services, and, first and foremost, the Maritime Prefect, who represents the state at sea, have launched several actions, including the implementation of measures from the Action Plan for the Marine Environment (*plan d'action pour le milieu marin*, PAMM), taken from the European Union's Marine Strategy Framework Directive (MSFD), in particular those relating to mooring.

Three measures of the PAMM are of specific concern to the Maritime Prefect, as he has been designated as their pilot. They relate to three categories of moorings for various vessels:

– the mooring of commercial vessels;

– the mooring of large yachts > 24 m (according to IMO classification, SOLAS convention);

– the mooring of small pleasure craft < 24 m.

The first measure, concerning the mooring of large pleasure craft, led to the issuance of Prefectoral Order 155/2016 of June 24, 2016.

Other measures are part of a mooring management strategy, the objective of which is the identification of sites at stake on the coastline of the seafront where the state can encourage the development of management solutions (installation of organized MLEZ[23] areas and mooring buoys or regulatory measures).

In this respect, the Maritime Prefecture has encouraged innovative projects that are able to meet not only marine safety objectives (buoying of the 300 m zone) but also environmental objectives (eco-designed buoys).

It should be noted that in the long term (within the framework of the objectives of the next PAMM cycle), only those infrastructures that are recognized as "ecological", that is, having little or no impact on the environment, will be authorized for all new installations. Bio-inspired structures, with a proven positive impact, could probably also be promoted.

Furthermore, aware of the environmental impact of government ships in operational use, the Navy has financed the installation of a mooring buoy in the Bay of Saint-Florent in Corsica, which will be followed by other similar installations on the mainland (Figure 4.56).

23 *Zones de mouillages et d'équipements légers pour la plaisance* (mooring and light equipment zones for pleasure craft).

Figure 4.56. *The Abeille Flandres moored in the Gulf of Saint-Florent in Corsica (photo: PREMAR)*

4.7.2.2. *Open water ecomoorings for large vessels and tourism yachts in Corsica*

The Mediterranean is the world's second largest cruising basin, and the distribution of capacities for yachts over 25 m favors the northern shore (Figure 4.57).

The main destinations are the North Coast of Sardinia, southern Corsica, the French Riviera, the Balearic Islands, the Amalfi Coast and the Gulf of Naples, the Aeolian Islands and the Croatian Islands. What these considerations highlight is a strong presence of large pleasure craft that is strongly at odds with natural ecosystems, leading to significant degradation, especially in the fragile and wild ecosystems of Corsica.

In order to overcome this problem, the Corsican authorities and the OEC aim to set up organized moorings for yachts all around Corsica.

The first of these projects, currently underway, will be located in Ajaccio and is steered by the CCI of Corsica which manages the Ajaccio Tino Rossi Marina.

A second project is emerging in Bonifacio and concerns about 15 mooring buoys for large yachts. This project is steered by the Bonifacio Marina with operational objectives from summer 2021.

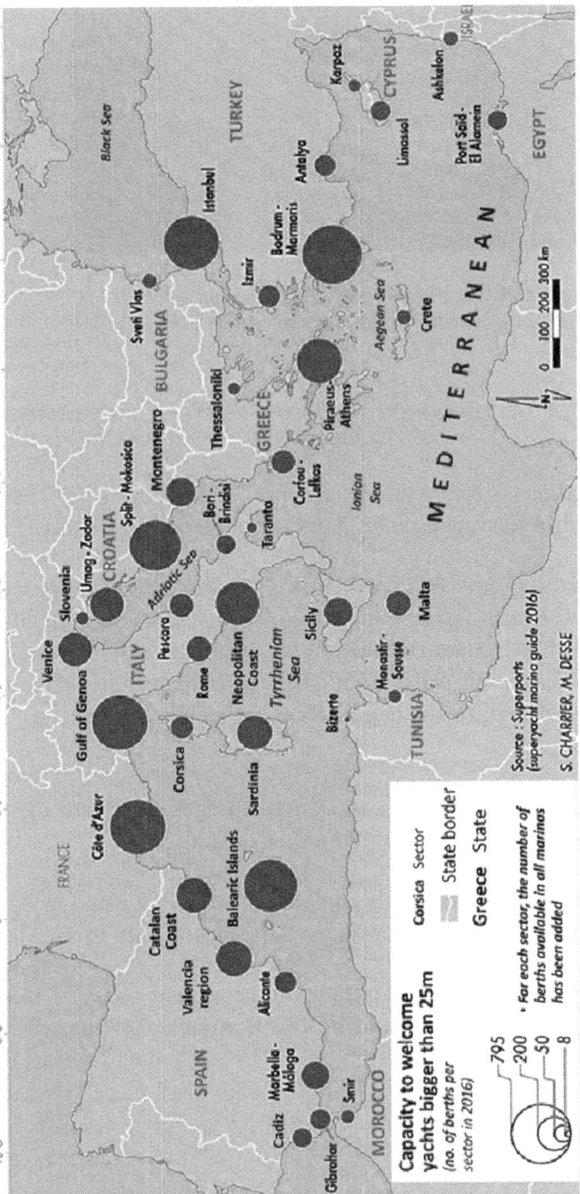

Figure 4.57. *Docking capacities for yachts over 25 m in the Mediterranean (Desse and Charrier 2017)*

Ajaccio's pilot project of organized moorings favors an eco-design approach. The weights are inspired by the shapes and surface conditions of local rocky habitats and target fish species[24]. Four moorings are planned in the gulf at Aspretto, Citadelle and Saint-François (Figure 4.58).

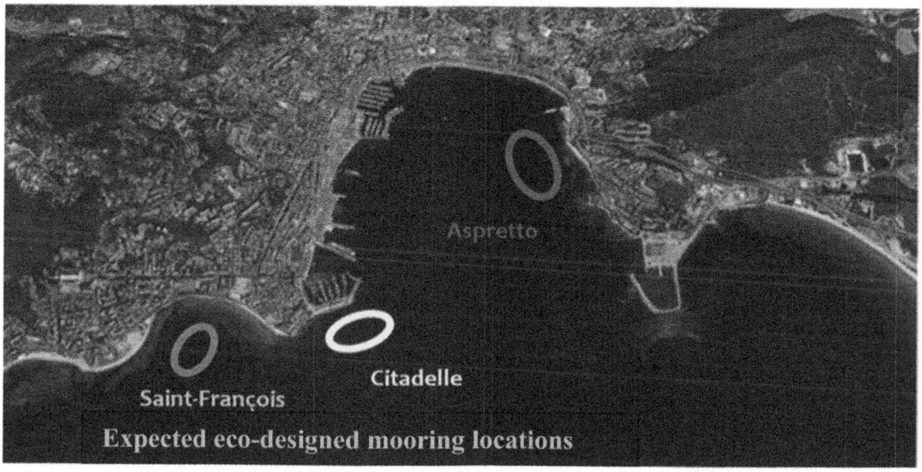

Figure 4.58. *Schematic layout of the planned mooring installation in the Gulf of Ajaccio, Corsica (according to CCI documents on a Google Earth background (map data, ©2021 Google))*

In this context, the company LIB Industries, associated with the engineering firm BRL Ingénierie, was contracted to manufacture one of the four buoys. With a mass of 34 tons, it is entirely eco-designed and adapted to the traction and resistance forces of large vessels (Figure 4.59). The structure is prefabricated and modular: this means that it can be transported in pieces by road and then assembled on the quayside before immersion, or it can even be assembled on the seabed after immersion using a team of hyperbaric divers.

The finished buoy weighs 34 t and is contained within an envelope 4.50 × 4.50 × 1.30 m in size and with a vacuum rate of over 40%. It is composed of a structural element that is adapted to take up the forces incurred by the ship and four corner elements with a combined weight contribution to the stability of the structure and spaces and habitats for local target species.

24 https://www.corsenetinfos.corsica/Plaisance-Le-Port-Tino-Rossi-d-Ajaccio-met-au-point-un-projet-pilote-d-amarrage-ecologique-pour-les-yachts_a34165.html.

Figure 4.59. *The 34 ton eco-designed mooring ballast manufactured by LIB Industries, AMarEco (photo: D. Debeauquenne, LIB Industries)*

The concept of the mechanical assembly of the elements is original (subject of a patent) and adapted to the prefabrication, transport and assembly on site of the buoys (saving time and resources). It favors a minimum of intervention time at sea in order to minimize the impact of works on the marine environment.

The surface of this structure was cast using elastomeric skins (LIB Industries) to mimic the texture of the local rock and give it a multi-scale roughness (Figure 4.60).

Biomimetic objectives are therefore the result of research on forms, the habitats created, the texture of surfaces and the materials used.

Figure 4.60. *Example of small concrete anchors for side marker buoys; photos taken after a season of immersion (manufacture: LIB Industries, AMarEco, photos: J.-L. Gaziello)*

"As someone who lives on the coast, in Lavandou, I can deplore the numerous boats at anchor in the Posidonia meadow. As a diver, I can only observe the multiple damage they cause as a result of anchoring. As a concrete manufacturer, I decided to participate in the implementation of alternative practices through the manufacture of concrete structures to meet the needs of organized moorings for pleasure boats or large yachts. The industrial tool of my company LIB Industries enables me to create these objects. However, prefabricated concrete moorings already exist for small tonnages.

The meeting between J.-C. Souche and S. Pioch was decisive since I understood that these structures could also be used as habitats for marine organisms provided that they are ecologically designed. The numerous discussions we had, including with professional fisherman Pierre Roy during the Ricantu Reef operation in Ajaccio, easily convinced me of the need to design and manufacture in a different way. At LIB Industries, we manufacture eco-designed, biomimetic concrete weights, because our objectives are to reproduce existing natural habitats near the structure to be built. Our mastery of elastomeric formwork enables us to reproduce almost any shape and texture of natural rock, using standard marine concrete, or even in future in biogenic concrete.

However, it is vital for the products manufactured that the cost be perfectly controlled, which has led us to design modular systems assembled from concrete wedges (patented process). It is therefore possible to manufacture large moorings in sections, then transport them to be assembled at the last moment on the immersion site, possibly at the bottom of the water.

In doing so, beyond the economic development of LIB in the marine field through the AMarEco brand, I hope that the response we provide to the societal and environmental need for moorings for pleasure boats and yachts will enable sensitive areas of the Mediterranean, such as Posidonia meadows and coralligenous areas, to be preserved."

Box 4.4. *Jean-Louis Gaziello's[25] experience, or why a concrete prefabrication manufacturer is converting to marine eco-design*

In addition to the operations relating to large-scale yachting, the OEC has endeavored to define a guide to the eco-design of harbor buoys or MLEZs for Corsican marinas. This action is part of a long-term strategic vision to orient all the island's marinas towards these new types of development (Figure 4.61).

Once again, the proposed structures meet environmental objectives and strong technical constraints: reduced draft in certain ports, highly variable seabeds (gravelly, sandy, muddy) and the various types of moorings envisaged by the different ports (by turning and/or by embossing).

25 President of LIB Industries.

Figure 4.61. *Photomontages of mooring system ready to be equipped (mooring line to surface buoy or main chain) (design: S. Pioch and J.-C. Souche)*

4.8. Eco-design of offshore viaduct piles

Located on Reunion Island, the viaduct of the new coastal road is the major structure of a project that also includes two sections of embankment road, which will replace the current Route de la Corniche linking Saint-Denis and La Possession and which is exposed to the risks of rock falls and submersion by the swell during cyclonic episodes. The project owner is the Reunion region.

The 5,400 m long sea viaduct and the sections of dike will link Saint-Denis (the capital of Reunion Island) to the Grande Chaloupe site (Figure 4.62). It will provide two three-lane highways, two lanes of which will be dedicated to soft modes of transport (buses, pedestrians and bicycles). The viaduct works amount to 715 million euro and are being built by a consortium comprising Vinci Construction Grands Projets, Vinci Construction, Bouygues Travaux Publics and Demathieu Bard.

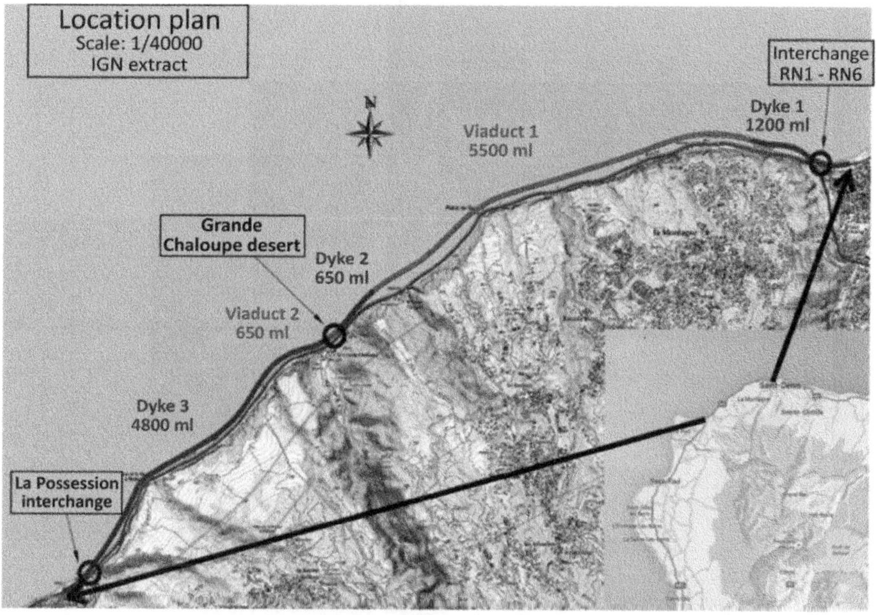

Figure 4.62. *New Coastal Road Project, Reunion Island, route and infrastructures on dikes (blue) and viaducts (red) over the sea (document: S. Pioch). For a color version of this figure, see www.iste.co.uk/pioch/marine.zip*

Usually, the piles of a viaduct are made of smooth concrete and are in a thick concrete footing resting on the seabed.

Eco-design studies therefore focused on the natural habitats and ethology of the coral species directly impacted by the project (sediments, noise, leveling of the seabed to lay the pile footings, etc.). Figures 4.63 and 4.64 illustrate the thought process for the biomimetic eco-design of the submerged sections of the engineering structure inspired by coral habitats.

The operational development of these proposals was carried out by the project management company Egis Eau, with several modifications.

Figure 4.63. *Eco-design of the medium part of the pillar (-5 to 10 m, colored in blue) with shapes mimicking naturally occurring coral table forms present near the structure (design: S. Pioch, drawing: J.-L. Feron). For a color version of this figure, see www.iste.co.uk/pioch/marine.zip*

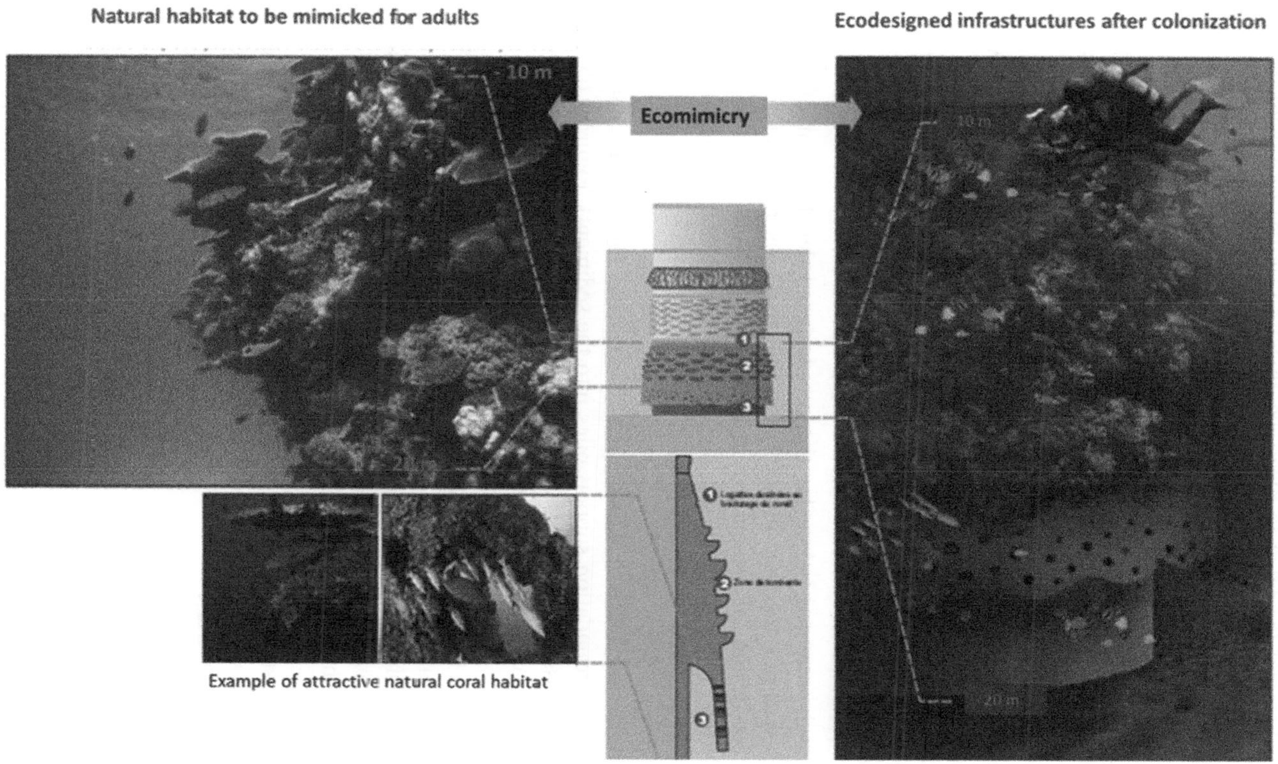

Figure 4.64. *Eco-design of the deeper part of the pillar (−10/20 m, colored in orange) with shapes mimicking naturally occurring coral falls and caves present near the structure (design: S. Pioch, drawing: J.-L. Feron).*
For a color version of this figure, see www.iste.co.uk/pioch/marine.zip

Even though the structure was not ultimately eco-designed *stricto sensu* since the structure itself was not designed to meet ecological objectives (the elements were added and fixed to the “classic” concrete of the piles), the idea of eco-efficient equipment fitted onto the viaduct piles was maintained. The work began at the end of 2014 and lasted six years.

4.9. Offshore wind farm project eco-design: multi-use perspectives

As part of its marine renewable energy development project, France has launched a vast program to design and build offshore wind turbines on the Channel and Atlantic coasts. The projects, which cover an area of about 100 km^2, involve not only foundations but also hundreds of kilometers of buried electrical cables linking each wind turbine and the landing cable connected to the terrestrial electrical grid. The area of seabed affected may have up to 150 ha destroyed.

The appraisal presented below for the eco-designed development of offshore wind turbines was put together in 2009. Its objective was to develop activities through the eco-design of structures, in order, on the one hand, to facilitate discussions during the consultation and, on the other hand, to reduce the unavoidable impacts of the offshore development project.

Wind turbines are often installed on soft seabeds, for ease of construction (drilling, driving of foundation support piles into the sediment). Figure 4.65 shows the layout of a “classic” wind turbine foundation and the attractive effect of the foundations on the flow of species, especially fish (larval and adult stages).

This “FAD” effect, named after these fish aggregating devices, lines connecting the seabed to the surface to attract pelagic species, is widely recognized as an ecological trap by fisheries scientists. It should be noted that in Japan, FADs are not considered as artificial reefs, let alone artificial habitats (in the sense described in Chapter 1), but as fishing gear (Pioch 2008).

The structures installed in the sea, which are disconnected from each other and offering little in the way of habitat and functions, are attractive because they are arranged on sandy “deserts”.

The risk is that they will constitute what scientific literature now refers to as “ecological traps”. Timothée Schwartz (Schwartz 2020) has warned of their deleterious effects.

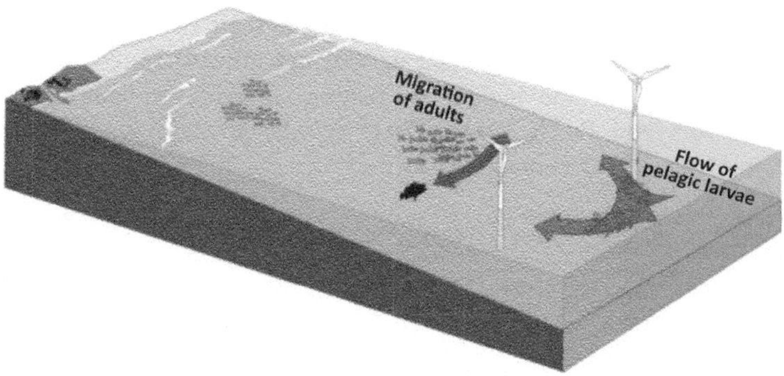

Figure 4.65. *Offshore wind turbine installation on soft seabeds and biological flows between foundation piles (design: S. Pioch, drawing: J.-L. Feron). For a color version of this figure, see www.iste.co.uk/pioch/marine.zip*

The objective, therefore, is to eco-design the foundations and the cables fixed to the seabed, in order to develop the available surfaces and functionalities for the colonization of species (fauna and flora) within an ecological framework designed so that the corridors increase resilience, as shown in Figure 4.66 in the deliberately simplifying and educational "before and after" style.

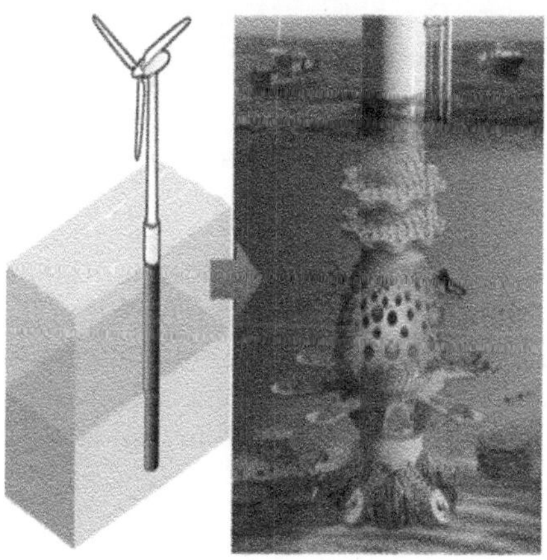

Figure 4.66. *Eco-design of offshore wind turbine foundations and cables: left, "classic" foundation; right, diagram of the idea of ecological functionalization and support for biodiversity, also ensuring bioprotection of the structure and a reduction in scouring (Lacroix and Pioch 2011). For a color version of this figure, see www.iste.co.uk/pioch/marine.zip*

The eco-design of these structures would enable their environmental integration to be improved, by using the important underwater infrastructures as a support for the ecosystems and for the benefit of fishing, tourism or nature conservation activities (Lacroix and Pioch 2011). These eco-design considerations thus fulfill multi-use objectives at sea (Figure 4.67).

Figure 4.67. *Ecological design of wind farms that also promotes multi-use through the ecological functions created (Lacroix and Pioch 2011)*

Ultimately, the concepts developed here were not implemented, as the studies showed that there was "no significant damage reported in the environmental impact study" and "a positive effect for the natural environment already provided by the foundations of usual design, as well as by the reservation of the area", without, however, any calculation making it possible to validate the assumed equivalence between the ecological losses and the gains.

Conclusion

The eco-design of marine structures is part of the design process of a development project. It is based on recent contributions in the areas of ecology, civil engineering and ecological engineering towards the desirable management and sustainable development of territories. This pragmatic approach attempts to respond to the research work that denounces the proven risks of introducing invasive species (Dafforn *et al.* 2009; Airoldi *et al.* 2015; Schaefer *et al.* 2020) and biodiversity loss (Bulleri and Chapman 2010) through "conventionally" designed coastal infrastructures. This approach also builds on promising experimental results from research teams from around the world, notably in Australia, England, Italy, the United States, Israel and France, which prove the effectiveness of nature-based solutions through bioprotection or the reduction of negative ecological impacts (La Marca *et al.* 2014; Dafforn *et al.* 2015; Pioch 2017). It is also one of the responses to the challenges of climate disruption, by improving the resilience of coastal areas. However, it is probably the recognition of the ecological gains, coupled with improved structural durability, for a controlled additional cost, that might, in the short term, seduce the stakeholders of coastal development (Iwanami *et al.* 2002; Airoldi *et al.* 2021). More broadly, the social demand for "greener" (or "blue") development projects, supported by policies advocating a necessary ecological transition, without net loss of biodiversity, will ensure the development of this new vision of development (Morris *et al.* 2019).

In practice, eco-design is still not widely known. It requires skills that are both transversal and deep. It also calls into question several traditional skills of the constructor who, in the interests of resistance and durability (over time), seeks simple forms, using smooth or antifouling materials and using very functional concepts. Now the trend is towards complex shapes, with rougher materials that become "soiled", due to the effect of significant biofouling, which is encouraged. The civil engineer has not been trained to take these additional environmental aspects into account. For their part, the environmental engineer is interested in biophysical principles, with an anticipatory eye on the environmental impacts of projects, but he has not been taught to dare to modify the very framework of a structure. The link could be eco-design, a marriage between civil engineering and environmental engineering, from which would be born a binomial design, resulting in resistant and durable (in both senses of the word) structures, favorable to biodiversity and, of course, at competitive costs.

Consequently, eco-design is the result of a desire to combine technology, ecology and innovation. In fact, it may not appeal to civil engineers, for whom it poses more problems than usual, or it may even be abused by operators lured by the prestige and tempted by the basic "eco-" marketing inherent in young sectors of activity, which will still be fragile in terms of the tools for controlling the cost/benefit efficiency ratio (ecological and socio-economic) of the various solutions promoted in the sometimes pompous name of "ecological engineering" or "nature-based solutions". These solutions can also be used as an excuse for building more and faster thanks to a green washing façade, a veritable Trojan horse for harmful development denounced by Firth *et al.* (2020).

We will leave it to time to sort out the ecological benefits that have been assessed and quantified for each of these innovations. This book, bringing together the experience of the two authors as well as numerous testimonies from scientists, structure owners, engineers and entrepreneurs, aims to stimulate creativity to facilitate the use of this integrated approach. Indeed, it gives a new meaning to the planner, whose choices can influence the development of marine life. Through all the testimonies presented, this book demonstrates that eco-design is a possible and realistic approach. We must seize it to do things differently and better.

To take this approach to its logical conclusion, eco-design can be an integral element of a sensitive approach, linked to a natural aesthetic that seems fundamental to us: the beautiful and bio-inspired. When the landscape and the aesthetics of a structure come together, a sense of harmony appears, the beautiful meets the good (McHarg 1969; Lassus 2002). The ecologist may be challenged about the futility of building a structure with an ecological function according to aesthetic "landscape" forms that mimic nature, when a cube with holes of the right dimensions would be suffice. The civil engineer may be familiar with a certain idea of "beauty" when working with a master architect. However, eco-design contains a sensitive approach, which would enable them to associate aestheticism, functionality and technicality. This sensitive aspect of development is fundamental in order to finally extract ourselves from reports only concerned with utility or composition of technical interests, revitalize the beauty that nature inspires in humankind and re-naturalize culture for a future full of hope. Thus, we could compare the vision of underwater development with musical notes. While the function of a note is to define a sound, harmony comes from organizing them aesthetically in the concert of the living. It is true that this vision may be accused of anthropocentrism at first sight, since it is the human view that verbalizes beauty. However, it can also be seen as a prelude to an attitude of respect for developing a "still natural" environment (Figure C.1). This is where the concept of bio-inspiration can help us. Indeed, although architectural beauty remains a subjective notion, which is specific to human culture, the fact of being inspired by nature, its landscapes, forms and ecological processes, whose efficiency has stood the test of time and the development of life, will guarantee that the eco-designed structure is "beautiful" through the effort of harmonious integration with its natural environment. This, for us, is the ultimate meaning of eco-design: to achieve functionally efficient, environmentally integrated, aesthetically "beautiful" developments; structures that we might think (looking at them from a distance) are part of the natural landscape. For this, new concepts, such as that of underwater landscape heritage[1], could be mobilized (Musard *et al.* 2007).

Figure C.1. *Photograph from the "Ecoreef" series, where an artificial reef was inspired by a natural aesthetic; a contemporary art, nature and society project (S. Pioch, black and white silver; composition: Nicolas le Floc'h, underwater artist[2])*

1 Inventory carried out in the framework of the Strategic Marine Framework Directive and cultural and aesthetic services.

2 http://www.nicolasfloch.net/.

The risk, however, would come from a tendency to sacralize or pantheize nature, so that the manager of an eco-designed structure is "trapped within the rules" when a protected species has settled and is no longer authorized to maintain or repair it. This is why it is a question of our relationship with nature, of "renaturalizing" our developmental thinking (Pelt 1977). In contrast, eco-design enables us to move away from the single logic of punitive environment, of environmental constraint, to pass beyond the game of "the developer who destroys then compensates". Humans are dependent on the environment, just as much as they are responsible for its unity and conservation. With eco-design, taking into account the natural environment becomes a source of inspiration and meaning in order to develop our creativity in the project of living on Earth.

Eco-design is an expertise, a sensitivity to nature, as much as a way of living intelligently with all living beings, for whom the most important task is to put life back at the heart of our human creations.

References

AFNOR (2014). NF EN 206/CN 2014. Concrete - Specification, performance, production and conformity - National addition to the standard NF EN 206. AFNOR, Seine-Saint-Denis.

Agropolis International (2014). Atelier de réflexion prospective MERMED. Adaptation aux changements globaux en mer Méditerranée.

Ahmed, K.B. (2016). Étude de l'encrassement biologique de matériaux cimentaires en eau de rivière: Analyse de l'influence des paramètres de surface des pâtes cimentaires. PhD thesis, University of Strasbourg, Strasbourg.

Airoldi, L. and Bulleri, F. (2011). Anthropogenic disturbance can determine the magnitude of opportunistic species responses on marine urban infrastructures. *PLoS One*, 6(8).

Airoldi, L., Beck, M.W., Firth, L.B., Bugnot, A.B., Steinberg, P.D., Dafforn, K.A. (2021). Emerging solutions to return nature to the urban ocean. *Annual Review of Marine Science*, 13, 445–477.

Airoldi, L., Turon, X., Perkol-Finkel, S., Rius, M. (2015). Corridors for aliens but not for natives: effects of marine urban sprawl at a regional scale. *Diversity and Distributions*, 21(7), 755–768.

Ammar, Y., Swailes, D., Bridgens, B., Chen, J. (2015). Influence of surface roughness on the initial formation of biofilm. *Surface and Coatings Technology*, 284, 410–416.

Anderson, M.J. and Underwood, A.J. (1994). Effects of substratum on the recruitment and development of an intertidal estuarine fouling assemblage. *Journal of Experimental Marine Biology and Ecology*, 184, 217–236.

Ardizzone, G.D., Gravina, M.F., Belluscio, A. (1989). Temporal development of epibenthic communities on artificial reefs in the central Mediterranean Sea. *Bulletin of Marine Science*, 44(2), 592–608.

Aronson, J., Clewell, A., Moreno-Mateos, D. (2016). Ecological restoration and ecological engineering: Complementary or indivisible? *Ecological Engineering*, 91, 392–395

Bailly, D. (1989). *Pêche et aquaculture au Japon*. Ifremer, Nantes.

Baine, M. (2001). Artificial reefs: A review of their design, application, management and performance. *Ocean and Coastal Management*, 44(3/4), 241–259.

Bajjouk, T., Guillaumont, B., Michez, N., Thouin, B., Croguennec, C., Populus, J., Louvel-Glaser, J., Gaudillat, V., Chevalier, C., Tourolle, J., Hamon, D. (2015). Classification EUNIS, système d'information européen sur la nature. Traduction française des habitats benthiques des régions Atlantique et Méditerranée. Vol. 1. Habitats Littoraux. Report, Ifremer, Nantes.

Baroghel-Bouny, V. (2004). Conception des bétons pour une durée de vie donnée des ouvrages – Maîtrise de la durabilité vis-à-vis de la corrosion des armatures et de l'alcali-réaction. Report, Association française du génie civil, Paris.

Belknap, R., Furtado, J., Forster, R., Blossom, H., Harris, C., Satterthwaite, A. (1967). *Three Approaches to Environmental Resource Analysis*. The Conservation Foundation, Washington.

Bellon-Fontaine, M., Benezech, T., Boutroux, K., Hermon, C. (2016). *Conception hygiénique de matériel et nettoyage-désinfection*. Lavoisier, Paris.

Béoutis, A., Jean, P., Colas, S. (2009). *L'observatoire du littoral: démographie et économie du littoral*. Insee, Montrouge.

Bergen, S.D., Bolton, S.M., Fridley, J.L. (2001). Design principles for ecological engineering. *Ecological Engineering*, 18(2), 201–210.

Bernier, G. (2004). Formulation des bétons, techniques de l'ingénieur. *Techniques de l'ingénieur*.

Berque, A. (1986). *Le sauvage et l'artifice: Les Japonais devant la nature*. Gallimard, Paris.

Berque, A. (1996). *Être humains sur la terre*. Gallimard, Paris.

Bigard, C., Pioch, S., Thompson, J.D. (2017). The inclusion of biodiversity in environmental impact assessment: Policy-related progress limited by gaps and semantic confusion. *Journal of Environmental Management*, 200, 35–45.

Bishop, M.J., Mayer-Pinto, M., Airoldi, L., Firth, L.B., Morris, R.L., Loke, L.H., Dafforn, K.A. (2017). Effects of ocean sprawl on ecological connectivity: Impacts and solutions. *Journal of Experimental Marine Biology and Ecology*, 492, 7–30.

Bonneuil, C. (2005). Les transformations des rapports entre sciences et société en France depuis la Seconde Guerre mondiale: un essai de synthèse. In *Actes du colloque Sciences, Médias et Société, École normale supérieure, lettres et sciences humaines*, Le Marec, J., Babou, I. (eds). ENS, Paris, 15–40.

Borsje, B.W., Van Wesenbeeck, B.K., Dekker, F., Paalvast, P., Bouma, T.J., Van Katwijk, M.M., de Vries, M., Bulleri, F. (2011). How ecological engineering can serve in coastal protection? *Ecological Engineering*, 37, 113–122.

Bouchon, C., Bouchon, Y., De Lavigne, S., Cordonnier, S. (2018). Suivi des communautés marines, zone de mouillages et d'équipements légers (ZMEL) en mouillages éco-récifs, commune de Deshaies. University of the Antilles, BOREA and Caraïbes Aqua Conseil.

Bouchoucha, M. (2017). Les zones portuaires peuvent-elles servir de nourriceries alternatives pour les poissons marins côtiers? Cas des sars en Méditerranée Nord-occidentale. PhD thesis, University of Perpignan, Perpignan.

Bourget, E., DeGuise, J., Daigle, G. (1994). Scales of substratum heterogeneity, structural complexity, and the early establishment of a marine epibenthic community. *Journal of Experimental Marine Biology and Ecology*, 181(1), 31–51.

Bugnot, A.B., Mayer-Pinto, M., Airoldi, L., Heery, E.C., Johnston, E.L., Critchley, L.P., Coleman, R.A., Sheehan, E.V. (2020). Current and projected global extent of marine built structures. *Nature Sustainability*, 4(1), 33–41

Bulleri, F. and Chapman, M.G. (2004). Intertidal assemblages on artificial and natural habitats in marinas on the north-west coast of Italy. *Marine Biology*, 145(2), 381–391.

Bulleri, F. and Chapman, M.G. (2010). The introduction of coastal infrastructure as a driver of change in marine environments. *Journal of Applied Ecology*, 47(1), 26–35.

Bulleri, F., Chapman, M.G., Underwood, A.J. (2005). Intertidal assemblages on seawalls and vertical rocky shores in Sydney Harbour (Australia). *Austral Ecology*, 30, 655–667.

Chapon, J. and Laval, M.-D. (1966). *Cours de l'École nationale des Ponts et Chaussées. Travaux maritimes*, volumes 1 and 2. Eyrolles, Paris.

Charbonnel, E., Serre, C., Ruitton, S., Harmelin, J. G., Jensen, A. (2002). Effects of increased habitat complexity on fish assemblages associated with large artificial reef units (French Mediterranean coast). *ICES Journal of Marine Science: Journal du Conseil*, 59(suppl.), S208–S213.

Chapman, M.G. and Underwood, A.J. (2011). Evaluation of ecological engineering of "armoured" shorelines to improve their value as habitat. *Journal of Experimental Marine Biology and Ecology*, 400(1), 302–313.

Charonnat, Y. (2012). Fabrication du béton hydraulique. *Techniques de l'ingénieur.*

Cheminée, A., Feunteun, E., Clerici, S., Cousin, B., Francour, P. (2014). Management of infralittoral habitats: Towards a seascape scale approach. In *Underwater Seascapes*, Musard, O., Le Dû-Blayo, L., Francour, P., Beurier, J.-P., Feunteun, E., Talassinos, L. (eds). Springer, Cham, 161–183.

Choi, C.G., Ohno, M., Sohn, C.H. (2006). Algal succession on different substrata covering the artificial iron reef at Ikata in Shikoku. *Japan Algae*, 21, 305–310.

Clavel, J., Julliard, R., Devictor, V. (2011). Worldwide decline of specialist species: Toward a global functional homogenization? *Frontiers in Ecology and the Environment*, 9(4), 222–228.

Clevenot, L., De Chastenet, C., Frascaria, N., Jacob, P., Raymond, R., Simon, L., Pech, P. (2017). Do linear transport infrastructures provide a potential corridor for urban biodiversity? Case study in Greater Paris, France. *Cybergeo: European Journal of Geography*. https://doi.org/10.4000/cybergeo.27895.

Cohen-Shacham, E., Walters, G., Janzen, C., Maginnis, S. (2016). *Nature-Based Solutions to Address Global Societal Challenges*. IUCN, Gland.

Connell, J.H. (1978). Diversity in tropical rain forests and coral reefs. *Science*, 199(4335), 1302–1310.

Connell, J.H. (1985). The consequences of variation in initial settlement vs. post-settlement mortality in rocky intertidal communities. *Journal of Experimental Marine Biology and Ecology*, 93(1/2), 11–45.

Connell, S.D. (2001). Urban structures as marine habitats: An experimental comparison of the composition and abundance of subtidal epibiota among pilings, pontoons and rocky reefs. *Marine Environmental Research*, 52(2), 115–125.

Cooke, S.J., Bergman, J.N., Nyboer, E.A., Reid, A.J., Gallagher, A.J., Hammerschlag, N., Vermaire, J.C. (2020). Overcoming the concrete conquest of aquatic ecosystems. *Biological Conservation*, 247(108589).

Coombes, M.A., Naylor, L.A., Thompson, R.C., Roast, S.D., Gomez-Pujol, L., Fairhurst, R.J. (2011). Colonization and weathering of engineering materials by marine microorganisms: An SEM study. *Earth Surface Processes and Landforms*, 36, 582–593.

Coombes, M.A., Naylor, L.A., Viles, H.A., Thompson, R.C. (2013). Bioprotection and disturbance: Seaweed, microclimatic stability and conditions for mechanical weathering in the intertidal zone [Online]. *Geomorphology*, 202, 4–14.

Coombes, M.A., La Marca, E.C., Naylor, L.A., Thompson, R.C. (2015). Getting into the groove: Opportunities to enhance the ecological value of hard coastal infrastructure using fine-scale surface textures. *Ecological Engineering*, 77, 314–323.

Coombes, M.A., Viles, H.A., Naylor, L.A., La Marca, E.C. (2017). Cool barnacles: Do common biogenic structures enhance or retard rates of deterioration of intertidal rocks and concrete. *Science of Total Environment*, 580, 1034–1045.

Crutzen, P.J. (2006). The "Anthropocene". In *Earth System Science in the Anthropocene*, Ehlers, E., Krafft, T. (eds). Springer, Berlin/Heidelberg, 13–18.

Dafforn, K.A., Johnston, E.L., Glasby, T.M. (2009). Shallow moving structures promote marine invader dominance. *Biofouling*, 25(3), 277–287.

Dafforn, K.A., Glasby, T.M., Johnston, E.L. (2012). Comparing the invasibility of experimental "reefs" with field observations of natural reefs and artificial structures. *PLoS One*, 7(5), e38124.

Dafforn, K.A., Glasby, T.M., Airoldi, L., Rivero, N.K., Mayer-Pinto, M., Johnston, E.L. (2015). Marine urbanization: An ecological framework for designing multifunctional artificial structures. *Frontiers in Ecology and the Environment*, 13(2), 82–90.

Dalod, E. (2015). Influence de la composition chimique de mortiers sur leur biodétérioration par les algues. PhD thesis, École nationale supérieure des Mines de Saint-Étienne, Saint-Étienne.

De Larrard, F. (2000). *Structures granulaires et formulation des bétons*. Laboratoire central des Ponts et Chaussées, Paris.

De Weerdt, K., Justnes, H., Geiker, M.R. (2014). Changes in the phase assemblage of concrete exposed to sea water. *Cement and Concrete Composites*, 47, 53–63.

Den Hartog, C. (1972). Substratum: Multicellular plants. *Marine Ecology*, 1, 1277–1289.

Dennis, H.D., Evans, A.J., Banner, A.J., Moore, P.J. (2018). Reefcrete: Reducing the environmental footprint of concretes for eco-engineering marine structures. *Ecological Engineering*, 120, 668–678.

Desse, M. and Charrier, S. (2017). La grande plaisance, un secteur économique en plein essor. *Études caribéennes*, 36, April.

Doody, J.P. (2004). Coastal squeeze. An historical perspective. *Journal of Coastal Conservation*, 10(1), 129–138.

Doumenge, F. (1961). Le Japon et l'exploitation de la mer. *L'information géographique*, 26(2), January–June, 91.

Doumenge, F. (1966). L'homme dans le Pacifique Sud. Étude géographique. *Revue française d'histoire d'outremer*, 53(192/193), 315–316.

Drouet, E. (2010). Impact de la température sur la carbonatation des matériaux cimentaires: Prise en compte des transferts hydriques. PhD thesis, École normale supérieure, Cachan.

Dublin, D.R. and Tanaka, N. (2014). Indigenous agricultural development for sustainability and "Satoyama". *Geography, Environment, Sustainability*, 7(2), 86–95.

Dubosc, A. (2000). Étude du développement de salissures biologiques sur les parements en béton: Mise au point d'essais accélérés de vieillissement. PhD thesis, INSA, Toulouse.

Dubosc, A., Escadeillas, G., Blanc, P.J. (2001). Characterization of biological stains on external concrete walls and influence of concrete as underlying material. *Cement and Concrete Research*, 31, 1613–1617.

El-Naggar, M.Y., Wanger, G., Leung, K.M., Yuzvinsky, T.D., Southam, G., Yang, J. (2010). Electrical transport along bacterial nanowires from *Shewanella oneidensis* MR-1. *Proceedings of the National Academy of Sciences*, 107, 18127–18131.

Evans, A.J., Firth, L.B., Hawkins, S.J., Morris, E.S., Goudge, H., Moore, P.J. (2016). Drill-cored rock pools: an effective method of ecological enhancement on artificial structures. *Marine and Freshwater Research*, 67(1), 123–130

Falque, M. (1972). Pour une planification écologique. *Options Méditerranéennes*, 13, 111–121.

Féral, F. (2011). L'extension récente de la taille des aires marines protégées: Une progression des surfaces inversement proportionnelle à leur normativité. *Vertigo*, special edition, 9.

Festa, J. and Dreux, G. (2007). *Nouveau guide du béton et de ses constituants*, 8th edition. Eyrolles, Paris.

Firth, L.B., Thompson, R.C., White, F.J., Schofield, M., Skov, M.W., Hoggart, S.P., Hawkins, S.J. (2013). The importance of water-retaining features for biodiversity on artificial intertidal coastal defence structures. *Diversity and Distributions*, 19(10), 1275–1283.

Firth, L.B., Schofield, M., White, F.J., Skov, M.W., Hawkins, S.J. (2014). Biodiversity in intertidal rock pools: Informing engineering criteria for artificial habitat enhancement in the built environment. *Marine Environmental Research*, 102, 122–130.

Firth, L.B., White, F.J., Schofield, M., Hanley, M.E., Burrows, M.T., Thompson, R.C., Skov M.W., Evans, A.J., Moore, P.J., Hawkins, S.J. (2016). Facing the future: the importance of substratum features for ecological engineering of artificial habitats in the rocky intertidal. *Marine and Freshwater Research*, 67(1), 131–143.

Firth, L.B., Airoldi, L., Bulleri, F., Challinor, S., Chee, S.Y., Evans, A.J., Hawkins, S.J. (2020). Greening of grey infrastructure should not be used as a Trojan horse to facilitate coastal development. *Journal of Applied Ecology*, 57, 1762–1768.

Fletcher, R.L. and Callow, M.E. (1992). The settlement, attachment and establishment of marine algal spores. *British Phycological Journal*, 27(3), 303–329.

Fourrier, A., Barral, M., De Montbrison, D., Jouvenel, J., Roche, E., Thorin, S., Pary, B. (2015). *Guide pratique d'aide à l'élaboration, l'exploitation et la gestion des récifs artificiels en Languedoc-Roussillon*. Cepralmar, Sète.

Francis, R.A. and Lorimer, J. (2011). Urban reconciliation ecology: The potential of living roofs and walls. *Journal of Environmental Management*, 92(6), 1429–1437.

Fritz-Feugeas, F., Cornet, A., Tribollet, B. (2008). *Biodétérioration des matériaux: Action des microorganismes, de l'échelle nanométrique à l'échelle macroscopique*. Ellipses, Paris.

Garcia, J. (2014). Influence des facteurs anthropiques et environnementaux sur les capacités de déplacements des poissons coralliens: Implications dans le design des aires marines protégées. PhD thesis, University of Perpignan, Perpignan.

Giannantonio, D.J., Kurth, J.C., Kurtis, K.E., Sobecky, P.A. (2009). Effects of concrete properties and nutrients on fungal colonization and fouling. *International Biodeterioration & Biodegradation*, 63, 252–259.

Glasby, T.M., Connell, S.D., Holloway, M.G., Hewitt, C.L. (2007). Nonindigenous biota on artificial structures: Could habitat creation facilitate biological invasions? *Marine Biology*, 151(3), 887–895.

Global Infrastructure Hub and Oxford Economics (2017). Global Infrastructure Outlook. Global Infrastructure Hub, Sydney.

Gobert, J. (2010). Éthique environnementale, remédiation écologique et compensations territoriales: entre antinomie et correspondances. *VertigO*, 10(1).

Gong, J., Song, W., Warren, A. (2005). Periphytic ciliate colonization: Annual cycle and responses to environmental conditions. *Aquatic Microbial Ecology*, 39, 159–170.

Govin, A., Tran, T.H., Guyonnet, R., Grosseau, P., Lors, C., Damidot, D. (2013). Ability in biofouling by *Klebsormidium flaccidum* of mortars: Influence of the intrinsic characteristics. *First International Conference on Concrete Sustainability (ICCS)*, S3-3(6), 914–919.

Green, R.H. (1979). *Sampling Design and Statistical Methods for Environmental Biologists*. Wiley, New York.

Green, D.S., Chapman, M.G., Blockley, D.J. (2012). Ecological consequences of the type of rock used in the construction of artificial boulder-fields. *Ecological Engineering*, 46, 1–10.

Guézennec, J. (2017). *Biodégration des matériaux: Quels risques pour la santé et l'environnement?* Quæ, Versailles.

Guillitte, O. (1995). Bioreceptivity: A new concept for building ecology studies. *Science of the Total Environment*, 167, 215–220.

Guillitte, O. and Dreesen, R. (1995). Laboratory chamber studies and petrographical analysis as bioreceptivity assessment tools of building materials. *Science of the Total Environment*, 167(1-3), 365–374.

Harmelin-Vivien, M.L., Harmelin, J.G., Leboulleux, V. (1995). Microhabitat requirements for settlement of juvenile sparid fishes on Mediterranean rocky shores. *Space Partition within Aquatic Ecosystems: Proceedings of the Second International Congress of Limnology and Oceanography*, Evian, May 25–28, 309–320.

Harshada, K., Desai, D., Gupta, A.K. (2017). Deterioration of concrete in marine structure. *Imperial Journal of Interdisciplinary Research*, 3(8).

Hayek, M., Salgues, M., Habouzlt, F., Bayle, S., Souche, J.C., De Weerdt, K., Pioch, S. (2020). *In vitro* and *in situ* tests to evaluate the bacterial colonization of cementitious materials in the marine environment. *Cement and Concrete Composites*, 113, 103748.

Hilbertz, W. (1978). Electrodeposition of minerals in seawater. *Oceans '78*, 699–706.

Hillier, S.R., Sangha, C.M., Plunkett, B.A., Walden, P.J. (1999). Long-term leaching of toxic trace metals from Portland cement concrete. *Cement and Concrete Research*, 29(4), 515–521.

Hughes, T.P., Kerry, J.T., Álvarez-Noriega, M., Álvarez-Romero, J.G., Anderson, K.D., Baird, A.H., Bridge, T.C. (2017). Global warming and recurrent mass bleaching of corals. *Nature*, 543(7645), 373–377.

Iacovidou, E. and Purnell, P. (2016). Mining the physical infrastructure: Opportunities, barriers and interventions in promoting structural components reuse. *Science of the Total Environment*, 557, 791–807.

IPBES (2019). Global assessment report on biodiversity and ecosystem services of the Intergovernmental. In *Science-Policy Platform on Biodiversity and Ecosystem Services*, Brondizio, E.S., Settele, J., Díaz, S., Ngo, H.T. (eds). IPBES secrétariat, Bonn.

Iwanami, M., Yokota, H., Hamada, H., Yamaji, T., Watanabe, H. (2002). Can marine fouling organisms extend the life of concrete structures? *Proceeding of the IABSE Symposium Towards a Better Built Environment–Innovation, Sustainability, Information Technology*, PAP164.

Jackson, A.C., Chapman, M.G., Underwood, A.J. (2008). Ecological interactions in the provision of habitat by urban development: Whelks and engineering by oysters on artificial seawalls. *Austral Ecology*, 33(3), 307–316.

Jacob, C., Pioch, S., Thorin, S. (2016). The effectiveness of the mitigation hierarchy in environmental impact studies on marine ecosystems: A case study in France. *Environmental Impact Assessment Review*, 60, 83–98.

Jensen, O. and Hansen, P. (2001). Water-entrained cement-based materials. I. Principles and theorical background. *Cem. Concr. Res.*, 31, 647–654.

Johnson, D.W. (1919). *Shore Processes and Shoreline Development*. Wiley, New York.

Kakimoto, H. (2004). *Design of Artificial Reef for Fisheries*, volume 2 (in Japanese). Japanese Institute for Promotion of Fisheries Ground and Communities, Tokyo.

Katsikogianni, M. and Missirlis, Y. (2004). Concise review of mechanisms of bacterial adhesion to biomaterials and of techniques used in estimating bacteria–material interactions. *European Cells and Materials*, 8, 37–57.

Kawabata, Y., Kato, E., Iwanami, M. (2012). Enhanced long-term resistance of concrete with marine sessile organisms to chloride ion penetration. *Journal of Advanced Concrete Technology*, 10(4), 151–159.

Kerr, A., Beveridge, C.M., Cowling, M.J., Hodgkiess, T., Parr, A.C.S., Smith, M.J. (1999). Some physical factors affecting the accumulation of biofouling. *Journal of the Marine Biological Association of the United Kingdom*, 79, 357–359.

Kool, J.T. and Nichol, S.L. (2015). Four-dimensional connectivity modelling with application to Australia's north and northwest marine environments. *Environmental Modelling and Software*, 65, 67–78.

Korbee, D., Mol, A.P., Van Tatenhove, J.P. (2014). Building with nature in marine infrastructure: Toward an innovative project arrangement in the Melbourne channel deepening project. *Coastal Management*, 42(1), 1–16.

Kuroki, T. (1952). The shape of a reef and its function as a habitat for fish. *Japan Fisheries Magazines*, 18(1).

Kurth, J.C., Krauss, P.D., Foster, S.W. (2019). Corrosion management of maritime infrastructure. *Transportation Research Record*, 2673(12), 2–14.

Lacovidou, E. and Purnell, P. (2016). Mining the physical infrastructure: Opportunities, barriers and interventions in promoting structural components reuse. *Sci. Total Environ.*, 557–558, 791–807.

Lacroix, D. and Pioch, S. (2011). The multi-use in wind farm projects: More conflicts or a win-win opportunity? *Aquatic Living Resources*, 24(2), 129–135.

Lacroix, D., Charbonnel, E., Mellon, C., Dao, J.C., Veron, G., Lagardere, JP., Coves, D., Buestel D. (2002). *Les récifs artificiels: Aménagement du littoral et repeuplement*. Ifremer, Paris.

La Marca, E.C., Coombes, M.A., Viles, H.A., Naylor, L.A. (2014). The bioprotective role of a biological encrustation. *Biologia Marina Mediterranea*, 21(1), 345–346.

Langlois, J., Junqa, G., Souche, J.C., Pioch, S. (2020). Environmental assessment of an offshore infrastructure using a beltline of concrete caissons and impact of logistical decisions. *Journées nationales Génie côtier – Génie civil. XVI*[es] *JNGCGC*. Le Havre, 8–10 December.

Lassus, P. (2002). *Harmonie et règle urbaine*. Anthropos, Paris.

Lee, H.S., Sidharthan, M., Shim, C.S., Kim, Y.D., Lim, C.Y., Ko, J.W., Han, M.D., Rang, M.J., Bim, L.S., Cho, H.S., Shin, H.W. (2008). Screening and formulation of chemoattractant coatings for artificial reef structures. *Journal of Environmental Biology*, 29, 605–612.

Leone, F. (2014). *Atlas des risques naturels et des vulnérabilités territoriales de Mayotte*. Presses universitaires de la Méditerranée, Montpellier.

Li, B., Reeve, D.E., Fleming, C.A. (2005). Design for enhanced marine habitats in coastal structures. *Proceedings of the Institution of Civil Engineers – Maritime Engineering*, 158(3), 115–122.

Li, K., Zhang, D., Li, Q., Fan, Z. (2019). Durability for concrete structures in marine environments of HZM project: Design, assessment and beyond. *Cement and Concrete Research*, 115, 545–558.

Lors, C., Feugeas, F., Tribollet, B. (2017). *Interactions matériaux-microorganismes: bétons et métaux plus résistants à la biodétérioration*. EDP Sciences, Paris.

Louisy, P. (2002). *Guide d'identification des poissons marins: Europe de l'Ouest et Méditerranée*. Eugen Ulmer, Paris.

Lukens, R. and Selberg, C. (2004). Guidelines for Marine Artificial Reef Materials [Online]. Available at: https://www.mbara.org/pdf/Guidelines_for_Marine_Artificial_Reef_Materials_January_2004.pdf.

Malyuk, V.V. (2019). Prognostication of concrete constructions durability based on experience in building and exploitation of maritime coast-protecting structures on Sakhalin Island. *MATEC Web of Conferences, EDP Sciences*, 265(02014).

Manso, S. and Aguado, A. (2016). The use of bio-receptive concrete as a new typology of living wall systems. *Matériaux & Techniques*, 104(5).

Manso, S., Mestres, G., Ginebra, M.P., De Belie, N., Segura, I., Aguado, A. (2014). Development of a low pH cementitious material to enlarge bioreceptivity. *Construction and Building Materials*, 54, 485–495.

Martins, G.M., Thompson, R.C., Neto, A.I., Hawkins, S.J., Jenkins, S.R. (2010). Enhancing stocks of the exploited limpet Patella candei d'Orbigny via modifications in coastal engineering. *Biological Conservation*, 143(1), 203–211.

Maruya, T., Iwanami, M., Sakai, E., Mashimo, M., Hamada, H. (2003). Durability enhancement of RC structures covered with a dense layer formed by marine fouling aquatic organisms. *Journal of Materials Concrete Structures Pavements*, 739(60), 61–74.

McHarg, I. (1969). *Design with Nature*. Natural History Press, New York.

McManus, R.S., Archibald, N., Comber, S., Knights, A.M., Thompson, R.C., Firth, L.B. (2018). Partial replacement of cement for waste aggregates in concrete coastal and marine infrastructure: A foundation for ecological enhancement? *Ecological Engineering*, 120, 655–667.

Meijer, J.R., Huijbregts, M.A., Schotten, K.C., Schipper, A.M. (2018). Global patterns of current and future road infrastructure. *Environmental Research Letters*, 13(6), 64006.

Menge, B.A., Ashkenas, L.R., Matson, A. (1983). Use of artificial holes in studying community development in cryptic marine habitats in a tropical rocky intertidal region. *Marine Biology*, 77, 129–142.

Miller, A.Z., Dionísio, A., Laiz, L., Macedo, M.F., Saiz-Jimenez, C. (2009). The influence of inherent properties of building limestones on their bioreceptivity to phototrophic microorganisms. *Annals of Microbiology*, 59, 705–713.

Miller, A.Z., Sanmartín, P., Pereira-Pardo, L., Dionísio, A., Saiz-Jimenez, C., Macedo, M.F., Prieto, B. (2012). Bioreceptivity of building stones: A review. *Sci. Total Environ*, 426 (1–12).

Mitsch, W.J. (2014). When will ecologists learn engineering and engineers learn ecology? *Ecological Engineering*, 65, 9–14.

Mitsch, W.J. and Jørgensen, S.E. (1989). Ecological engineering: An introduction to eco-technology [Online]. Available at: https://www.osti.gov/biblio/6281717-ecological-engineering-introduction-ecotechnology.

Morris, R.L., Heery, E.C., Loke, L.H., Lau, E., Strain, E., Airoldi, L., Dong, Y.W. (2019). H4 design options, implementation issues and evaluating success of ecologically engineered shorelines. *Oceanography and Marine Biology: An Annual Review*, 57, 169–228.

Mos, B., Dworjanyn, S.A., Mamo, L.T., Kelaher, B.P. (2019). Building global change resilience: Concrete has the potential to ameliorate the negative effects of climate-driven ocean change on a newly-settled calcifying invertebrate. *Science of the Total Environment*, 646, 1349–1358.

Moschella, P.S., Abbiati, M., Åberg, P., Airoldi, L., Anderson, J.M., Bacchiocchi, F., Granhag, L. (2005). Low-crested coastal defence structures as artificial habitats for marine life: using ecological criteria in design. *Coastal Engineering*, 52(10/11), 1053–1071.

Mundy, C.N. and Babcock, R.C. (1998). Role of light intensity and spectral quality in coral settlement: Implications for depth-dependent settlement? *Journal of Experimental Marine Biology and Ecology*, 223, 235–255.

Musard, O., Fournier, J., Marchand, J.P. (2007). The underwater space: Essay on the concept of landscape. *L'Espace geographique*, 36(2), 168–185.

Nakamura, M. (1985). Evolution of artificial fishing reef concepts in Japan. *Bulletin of Marine Science*, 37, 271–278.

Nickels, J.S., Bobbie, R.J., Lott, D.F., Martz, R.F., Benson, P.H., White, D.C. (1981). Effect of manual brush cleaning on biomass and community structure of micro-fouling film formed on aluminum and titanium surfaces exposed to rapidly flowing seawater. *Applied and Environmental Microbiology*, 41(6), 1442–1453.

Nuhoglu, Y., Oguz, E., Uslu, H., Ozbek, A., Ipekoglu, B., Ocak, I., Hasenekoglu, I. (2006). The accelerating effects of the microorganisms on biodeterioration of stone monuments under air pollution and continental-cold climatic conditions in Erzurum, Turkey. *Science of the Total Environment*, 364, 272–283.

Odum, H.T., Siler, W.L., Beyers, R.J., Armstrong, N. (1963). Experiments with engineering of marine ecosystems. *Publications of the Institute of Marine Science, University of Texas*, 9, 374–403.

Odum, H.T., Copeland, B.J., McMahan, E.A. (1974). *Coastal Ecological Systems of the United States*, volume 4. The Conservation Foundation, Washington.

Office français de la Biodiversité (2018). Guide d'élaboration des plans de gestion des espaces naturels [Online]. Available at: http://ct88.espaces-naturels.fr/.

Pastor, J. (2008). Rôle des enrochements côtiers artificiels dans la connectivité des populations, cas du sar commun (*Diplodus sargus*, Linné, 1758) en Méditerranée nord-occidentale. PhD thesis, University of Perpignan, Perpignan.

Patranella, A., Kilfoyle, K., Pioch, S., Spieler, R.E. (2017). Artificial reefs as juvenile fish habitat in a marina. *Journal of Coastal Research*, 33(6), 1341–1351.

Pelt, J.-M. (1977). *L'homme re-naturé*. Le Seuil, Paris.

Pérès, C. (2006). Synthèse des types de corps-morts permanents pour le mouillage des bateaux de plongée sur les côtes de Martinique. Observatoire du milieu marin martiniquais. Report, OMMM.

Perkol-Finkel, S. and Sella, I. (2014). Ecologically active concrete for coastal and marine infrastructure: Innovative matrices and designs. In *From Sea to Shore? Meeting the Challenges of the Sea*, Allsop, W, Burgess, K. (eds). ICE Publishing, London, 1139–1149.

Perkol-Finkel, S., Hadary, T., Rella, A., Shirazi, R., Sella, I. (2018). Seascape architecture – incorporating ecological considerations in design of coastal and marine infrastructure. *Ecological Engineering*, 120, 645–654.

Persoone, G. (1971). Ecology of fouling on submerged surfaces in a polluted harbour. *Vie et Milieu*, 22, 613–636.

Phan, K.L., Van Thiel de Vries, J.S., Stive, M.J.F. (2014). Coastal mangrove squeeze in the Mekong Delta. *Journal of Coastal Research*, 31(2), 233–243.

Powers, T. and Bromnyard, T. (1946). Studies of the physical properties of hardened portland Cement Paste. *J. Proc.*, 43, 101–132.

PIANC (2011a). Towards a sustainable waterborne transportation industry [Online]. EnviCom Task Group 2. Available at: https://www.pianc.org/publications/envicom/towards-a-sustainable-waterborne-transportation-industry.

PIANC (2011b). Working with Nature [Online]. PIANC position paper. Available at: https://www.pianc.org/uploads/files/EnviCom/WwN/WwN-Position-Paper-English.pdf.

PIANC (2019). A practical guide to environmental risk management (ERM) for navigation infrastructure projects [Online]. EnviCom WG 175. Available at: https://www.pianc.org/publications/envicom/wg175.

PIANC (2020). Climate change adaptation planning for ports and inland waterways [Online]. EnviCom WG 178. Available at: https://www.pianc.org/publications/envicom/wg178.

Pinault, M., Pioch, S., Nicolas, P. (2017). Guide pour les études d'impact environnemental (EIE) en milieu corallien de France et d'outremer [Online]. Ifrecor. Available at: http://www.ifrecor-doc.fr/files/original/ef79d2f1270e585d9fb8a57d80bce76b.pdf.

Pinot, J.-P. (1998). *La gestion du littoral*. Institut océanographique, Paris.

Pioch, S. (2008). Les "habitats artificiels": Elément de stratégie pour une gestion intégrée des zones côtières? Essai de méthodologie d'aménagement en récifs artificiels adaptés à la pêche artisanale côtière. PhD thesis, Paul Valéry University Montpellier 3, Montpellier.

Pioch, S. (2017). Vers une nouvelle gouvernance côtière entre aménagement et environnement? La compensation des impacts de l'homme sur l'environnement dans les projets d'aménagements maritimes. Thesis, Paul Valéry University Montpellier 3, Montpellier.

Pioch, S. and Léocadie, A. (2017). Overview on eco-moorings facilities: Commented bibliography [Online]. International Coral Reef Initiative, Foundation for Research on Biodiversity report. Available at: https://www.icriforum.org/wp-content/uploads/2019/12/OVERVIEW%20of%20eco-mooring-light.pdf [Accessed 28 December 2020].

Pioch, S., Levrel, H., Hay, J., Frascaria-Lacoste, N., Martin, G. (2015). *Restaurer la nature pour atténuer les impacts du développement: Analyse des mesures compensatoires pour la biodiversité*. Quæ, Versailles.

Pioch, S., Relini, G., Souche, J.C., Stive, M.J.F., De Monbrison, D., Nassif, S., Kilfoyle, K. (2018). Enhancing eco-engineering of coastal infrastructure with eco-design: Moving from mitigation to integration. *Ecological Engineering*, 120, 574–584.

Ramade, F. (2009). *Éléments d'écologie. Écologie fondamentale*, 4th edition. Dunod, Paris.

Reaka-Kudla, M.L. (1997). The global biodiversity of coral reefs: a comparison with rain forests. In *Biodiversity II: Understanding and Protecting our Biological Resources*, Wilson, D.E., Reaka-Kudla, M.L. (eds). Joseph Henry Press, Washington.

Recybeton (2012). Complete recycling of concrete [Online]. Available at: http://www.pnrecybeton.fr [Accessed 30 December 2020].

Relini, G. (1993). Mediterranean macrofouling. *Oebalia*, 19(Suppl.), 103–154.

Rice, J. C. (2005). Understanding fish habitat ecology to achieve conservation. *Journal of Fish Biology*, 67, 1–22.

Roberts, D.J., Nica, D., Zuo, G., Davis, J.L. (2002). Quantifying microbially induced deterioration of concrete: initial studies. *International Biodeterioration & Biodegradation*, 49, 227–234.

Rosenzweig, M.L. and Michael, L. (2003). *Win-Win Ecology: How the Earth's Species can Survive in the Midst of Human Enterprise*. Oxford University Press, Oxford.

Roux, S., Feugeas, F., Cornet, A. (2006). Altération des pâtes de ciment par colonisation bactérienne. *Matériaux & Techniques*, 94, 495–506.

Ruhl, J.B. and Craig, R.K. (2009). New sustainable governance institutions for estuaries and coasts. Treatise on estuarine and coastal science [Online]. FSU, College of Law, Public Law Research. Available at: https://ssrn.com/abstract=1460981.

Ruitton, S. (1999). Les communautés benthiques et nectobenthiques associées aux aménagements littoraux en Méditerranée nord-occidentale: structure et fonctionnement. PhD thesis, University of the Mediterranean Aix-Marseille II, Aix-en-Provence.

Rytwinski, T., Soanes, K., Jaeger, J.A., Fahrig, L., Findlay, C.S., Houlahan, J., Van der Grift, E.A. (2016). How effective is road mitigation at reducing road-kill? A meta-analysis. *PLoS One*, 11(11), e0166941.

Salgues, M. (2013). Modélisation des effets structuraux des réactions sulfatiques internes : application aux barrages en béton. PhD thesis.

Salgues, M., Pioch, S., Souche, J.-C., De Weerdt, K. (2020). L'écoconception maritime: Une révolution pour les maîtres d'ouvrages. *Revue Paralia*, 13.

Schaefer, N., Bishop, M.J., Bugnot, A.B., Foster-Thorpe, C., Herbert, B., Hoey, A., Mayer-Pinto, M., Sherman, C.D.H., Vozzo, M.L., Dafforn, K.A. (2020). Ecological engineering of marine infrastructure for biosecurity [Online]. Available at: https://www.researchgate.net/profile/Katherine_Dafforn/publication/343760709_Ecological_engineering_of_marine_infrastructure_for_biosecurity_Report_prepared_for_Department_of_Agriculture_Water_and_the_Environment/links/5f3f635f92851cd3020edc75/Ecological-engineering-of-marine-infrastructure-for-biosecurity-Report-prepared-for-Department-of-Agriculture-Water-and-the-Environment.pdf.

Schwartz, T. (2020). Les dispositifs artificiels au service de la restauration et de la compensation écologique: de l'évaluation du risque de piège écologique aux recommandations de bonnes pratiques. PhD thesis, Paris Sciences et Lettres University/l'École pratique des hautes études, Paris.

Seaman Jr, W. and Sprague, L.M. (eds) (1991). Artificial habitat practices in aquatic systems. *Artificial Habitats for Marine and Freshwater Fisheries*. Academic Press, Cambridge, 1–29.

Sella, I. and Perkol-Finkel, S. (2015). Blue is the new green: Ecological enhancement of concrete-based coastal and marine infrastructure. *Ecological Engineering*, 84, 260–272.

Seto, K.C., Güneralp, B., Hutyra, L.R. (2012). Global forecasts of urban expansion to 2030 and direct impacts on biodiversity and carbon pools. *Proceedings of the National Academy of Sciences*, 109(40), 16083–16088.

Sheng, Y.P. (2000). Physical characteristics and engineering at reef. In *Artificial Reef Evaluation with Application to Natural Marine Habitats*, Seaman, W. (ed.). CRC Press, Boca Raton, 51–94.

Shepard, F.P. (1965). *Submarine Geology*, 2nd edition. Literary Licensing, New York.

Sherman, R.L., Gilliam, D.S., Spieler, R.E. (2001). Site-dependent differences in artificial reef function: Implications for coral reef restoration. *Bulletin of Marine Science*, 69, 1053–1056.

Shi, X. and Zhu, X. (2009). Biofilm formation and food safety in food industries. *Trends in Food Science & Technology*, 20, 407–413.

Simard, F. (1996). Socio-economic aspects of artificial reefs in Japan. *Proceeding of the first EARRN Conference*, Ancône.

Somsueb, S., Ohno, M., Kimura, H. (2001). Development of seaweed communities on suspended substrata with three slope angles. *Journal of Applied Phycology*, 13, 109–115.

Souche, J.-C., Le Saout, G., Salgues, M., Pioch, S. (2016). Effets de bétons bioactifs sur la colonisation marine en environnement méditerranéen. *Matériaux & Techniques*, 104, 504.

Souche, J.-C., Pioch, S., Salgues, M., Ienny, P., Le Saout, G. (2018a). *Interaction des ouvrages avec leur environnement: Le milieu maritime*. Presses des Mines, Paris.

Souche, J.-C., Pioch, S., Salgues, M., De Weerdt, K. (2018b). L'écoconception maritime: Le paramètre matériaux. *Conférence internationale francophone NoMaD*. University of Liège, Liège.

Souche, J.-C., Pioch, S., Salgues, M., De Weerdt, K., Agostini, A., Hayek, M. (2019). De la conception à l'écoconception des ouvrages maritimes: intégrer la nature au projet d'aménagement maritime. *Paralia*, 12.

Steinberg, P.D., Airoldi, L., Banks, J., Leung, K.M. (2016). Introduction to the special issue on the World Harbour Project. *Regional Studies in Marine Science*, 8, 217–219.

Taheri, S. (2019). A review on five key sensors for monitoring of concrete structures. *Construction and Building Materials*, 204, 492–509.

Tanoue, H., Miyazaki, N., Niizawa, T., Mizushima, K., Suzuki, M., Ruitton, S., Biton Porsmoguer, S., Alabsi, N., Gonzalvo, S., Mohri, M., Hamano, A., Komatsu, T. (2015). Measurement of fish habitat use by fish-mounted data loggers for integrated coastal management: An example of Japanese sea bass (*Lateolabrax japonicus*). In *Marine Productivity: Perturbations and Resilience of Socio-ecosystems*, Ceccaldi, H.-J., Hénocque, Y., Koike, Y., Komatsu, T., Stora, G., Tusseau-Vuillemin, M.-H. (eds). Springer International Publishing, New York, 243–251.

Tarlet, J. (1977). Milieu naturel et aménagement. Les méthodes de planification écologique. *Annales de géographie*, 86(474), 164–200.

Terlizzi, A. and Faimali, M. (2010). Fouling on artificial substrata. In *Biofouling*. Dürr, S., Thomason, J. (eds), Wiley-Blackwell, London, 170–184.

Teughels, W., Assche, N.V., Sliepen, I., Quirynen, M. (2006). Effect of material characteristics and/or surface topography on biofilm development. *Clinical Oral Implants Research*, 17, 68–81.

Tomaselli, L., Lamenti, G., Bosco, M., Tiano, P. (2000) Biodiversity of photosynthetic microorganisms dwelling on stone monuments. *International Biodeterioration & Biodegradation*, 46, 251–258.

Tran, T.H. (2011). Influence of the intrinsic characteristics of a mortar on its ability to resist to the biofouling [Online]. Available at: https://tel.archives-ouvertes.fr/tel-00718250 [Accessed 28 December 2020].

Tran, T.H., Govin, A., Guyonnet, R., Grosseau, P., Lors, C., Garcia-Diaz, E., Damidot, D., Devès, O., Ruot, B. (2012). Influence of the intrinsic characteristics of mortars on biofouling by Klebsormidium flaccidum. *International Biodeterioration & Biodegradation*, 70, 31–39.

Tran, T.H., Govin, A., Guyonnet, R., Grosseau, P., Lors, C., Damidot, D. (2014). Influence of the intrinsic characteristics of mortars on their biofouling by pigmented organisms: Comparison between laboratory and field-scale experiments. *International Biodeterioration and Biodegradation*, 86, 334–342.

Valdora, P.F., Gomez, A.G., Steinberg, P. *et al.* (2020). A global approach to mapping the environmental risk of harbours on aquatic systems. *Marine Policy*, 119, 104051.

Van Bohemen, H.D. (2004). Ecological engineering and civil engineering works. PhD thesis, Delft University of Technology, Delft.

Vaz-Pinto, F., Torrontegi, O., Prestes, A.C.L., Álvaro, N.V., Neto, A.I., Martins, G.M. (2014). Invasion success and development of benthic assemblages: effect of timing, duration of submersion and substrate type. *Marine Environmental Research*, 94, 72–79.

Vincke, E., Boon, N., Verstraete, W. (2001) Analysis of the microbial communities on corroded concrete sewer pipes-a case study. *Applied Microbiology and Biotechnology*, 57, 776–785.

Waltham, N.J. and Sheaves, M. (2018). Eco-engineering rock pools to a seawall in a tropical estuary: Microhabitat features and fine sediment accumulation. *Ecological Engineering*, 120, 631–636.

Whitman, W.B., Coleman, D.C., Wiebe, W.J. (1998). Prokaryotes: The unseen majority. *Proceedings of the National Academy of Sciences*, 95, 6578–6583.

Wilding, T.A. and Sayer, M.D.J. (2002). The physical and chemical performance of artificial reef blocks made using quarry by-products. *ICES Journal of Marine Science*, 59(Suppl.), 250–257.

Xu, Q., Ji, T., Yang, Z., Ye, Y. (2019). Preliminary investigation of artificial reef concrete with sulphoaluminate cement, Marine sand and sea water. *Construction and Building Materials*, 211, 837–846.

Yamine, J. (2007). Rhéologie des bétons fluides à hautes performances: Relations entre formulations, propriétés rhéologiques, physicochimie et propriétés mécaniques. PhD thesis, École normale supérieure de Cachan, Gif-sur-Yvette.

Yanagi, T. (2012). *Japanese Commons in the Coastal Seas: How the Satoumi Concept Harmonizes Human Activity in Coastal Seas with High Productivity and Diversity*. Springer Science & Business Media, Berlin.

Zobell, C.E. (1972). Substratum: Bacteria, fungi and blue-green algae. *Marine Ecology*, 1(3), 1251–1270.

Index

A, B, C

D, E

H, I, M

N, P, R

S, T

U, W

Other titles from

iSTE

in

Ecological Science

2021

AMIARD Jean-Claude
Disarmament and Decommissioning in the Nuclear Domain (Radioactive Risk SET – Volume 4)

JOLY Fernand, BOURRIÉ Guilhem
Mankind and Deserts 1: Deserts, Aridity, Exploration and Conquests
Mankind and Deserts 2: Water and Salts
Mankind and Deserts 3: Wind in Deserts and Civilizations

MUTTIN Frédéric, THOMAS Hélène
Marine Environmental Quality

2020

BRUSLÉ Jacques, QUIGNARD Jean-Pierre
Fish Behavior 1: Eco-ethology
Fish Behavior 2: Ethophysiology

LE FLOCH Stéphane
Remote Detection and Maritime Pollution: Chemical Spill Studies

MIGON Christophe, NIVAL Paul, SCIANDRA Antoine
The Mediterranean Sea in the Era of Global Change 1: 30 Years of Multidisciplinary Study of the Ligurian Sea
The Mediterranean Sea in the Era of Global Change 2: 30 Years of Multidisciplinary Study of the Ligurian Sea

ROSSIGNOL Jean-Yves
Climatic Impact of Activities: Methodological Guide for Analysis and Action

2019

AMIARD Jean-Claude
Industrial and Medical Nuclear Accidents: Environmental, Ecological, Health and Socio-economic Consequences
(Radioactive Risk SET – Volume 2)
Nuclear Accidents: Prevention and Management of an Accidental Crisis
(Radioactive Risk SET – Volume 3)

BOULEAU Gabrielle
Politicization of Ecological Issues: From Environmental Forms to Environmental Motives

DAVID Valérie
Statistics in Environmental Sciences

GIRAULT Yves
UNESCO Global Geoparks: Tension Between Territorial Development and Heritage Enhancement

KARA Mohamed Hichem, QUIGNARD Jean-Pierre
Fishes in Lagoons and Estuaries in the Mediterranean 2: Sedentary Fish
Fishes in Lagoons and Estuaries in the Mediterranean 3A: Migratory Fish
Fishes in Lagoons and Estuaries in the Mediterranean 3B: Migratory Fish

OUVRARD Benjamin, STENGER Anne
Incentives and Environmental Policies: From Theory to Empirical Novelties

2018

AMIARD Jean-Claude
Military Nuclear Accidents: Environmental, Ecological, Health and Socio-economic Consequences
(Radioactive Risk SET – Volume 1)

FLIPO Fabrice
The Coming Authoritarian Ecology

GUILLOUX Bleuenn
Marine Genetic Resources, R&D and the Law 1: Complex Objects of Use

KARA Mohamed Hichem, QUIGNARD Jean-Pierre
Fishes in Lagoons and Estuaries in the Mediterranean 1: Diversity, Bioecology and Exploitation

2016

BAGNÈRES Anne-Geneviève, HOSSAERT-MCKEY Martine
Chemical Ecology

2014

DE LARMINAT Philippe
Climate Change: Identification and Projections